Budgerigar

Budgerigar

How a brave, chatty and colourful little
Aussie bird stole the world's heart

SARAH HARRIS & DON BAKER

ALLEN&UNWIN
SYDNEY·MELBOURNE·AUCKLAND·LONDON

First published in 2020

Allen & Unwin
83 Alexander Street
Crows Nest NSW 2065
Australia
Phone: (61 2) 8425 0100
Email: info@allenandunwin.com
Web: www.allenandunwin.com

A catalogue record for this
book is available from the
National Library of Australia

ISBN 978 1 76087 548 0

Illustrations by Mika Tabata
Set in 12/16 pt Minion Pro by Midland Typesetters, Australia
Printed and bound in Australia by Griffin Press

10 9 8 7 6 5 4 3 2 1

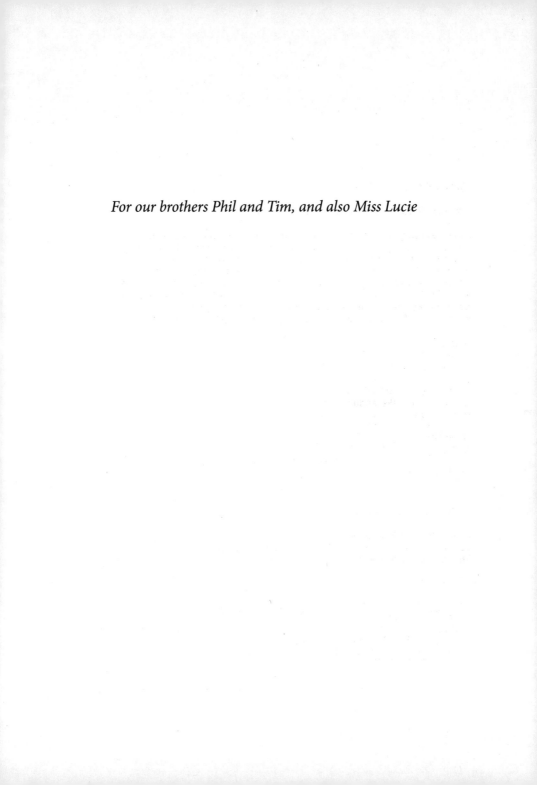

For our brothers Phil and Tim, and also Miss Lucie

Contents

Budgerigar

A companion to the great

SIR WINSTON CHURCHILL sat up in his bed, ubiquitous unlit cigar hanging fatly from his lip and a small blue-green bird perched atop his head, and dictated to his secretary. Between occasional swooping sips of his master's whisky and soda, Toby the budgie dropped his own little bombs on the top-secret thermonuclear project documents spread across the British prime minister's counterpane. Defence minister Harold Macmillan recalled the scene in his diary entry for 26 January 1955: 'this little bird flew around the room, perched on my shoulder and pecked (or kissed my cheek) … while sonorous sentences were rolling out of the maestro's mouth on the most terrible and destructive engine of mass warfare yet known to mankind. The bird says a few words in a husky voice like an American actress.'

At age eleven, Richard Branson's first enterprise—growing Christmas trees—was nibbled into non-existence by rabbits. Undeterred, he and best mate Nik Powell hit on a scheme to become budgie barons, with Richard convincing his father to build a large aviary on the family's Surrey property. But the budgies established themselves rather too successfully. 'Even after everyone in Shamley Green had bought at least two, we were still left with an aviary full of them,' the music, technology and transport magnate recalled in his autobiography *Losing My Virginity*. Not long after the boys returned to boarding school, Richard received a letter from his mother lamenting that rats had invaded the aviary and eaten all the budgies. It was many years before she admitted she had actually left the door open and they'd all escaped.

In the late 1920s, the Japanese nobility, led by Emperor Hirohito, ignited an extraordinary boom in the budgerigar market when the birds became popular as betrothal gifts from families of wealthy grooms to their brides-to-be. Hirohito, a student of natural history, had been enchanted by budgerigars on a 1921 visit to Europe before his ascension to the Chrysanthemum Throne. He later said it had been during this trip that 'I, who had been a bird in a cage, first experienced freedom'. In 1927 a member of the Imperial household paid the princely sum of £175 (the equivalent of more than $13,000 today) for a single blue budgie. But when there was a global outbreak of psittacosis (parrot fever) in 1929–30, the market crashed and for a time the international trade in budgerigars was embargoed.

Chapter 1

Unbeknown to early tourists

IT IS GENERALLY supposed that the budgie didn't make it onto the ark of dead things Captain James Cook took back to England with him in 1771 after his voyage of discovery to find Terra Australis. But that was not for want of trying by the feverish naturalists who had accompanied him.

Any creature within their sights became fair game—quite literally. The natural history artist Sydney Parkinson described the *Endeavour*'s entry into Botany Bay thus: '[There were] a great number of birds of a beautiful plumage; among which were two sorts of parroquets, and a beautiful loriquet: we shot a few of them, which we made into a pie, and they ate very well. We also met with a black bird, very much like our crow, and shot some of them too, which also tasted agreeably.'

But while he observed, and indeed ate, any number of birds, Parkinson only actually drew one of them. This reflects the expedition's extreme bias towards botany, a bias shared by its chief scientist Joseph Banks, who later enthused to his aristocratic French friend and chemist, the Comte de Lauraguais: 'The Number of Natural productions discover'd in this Voyage is incredible: about 1000 Species of Plants that have not been at all describ'd by any Botanical author.' The '500 fish, as many Birds, and insects [of the] Sea & Land innumerable' were very much also-rans.

While Banks scrupulously kept and catalogued the botanical material he'd gathered in New Holland, the zoological specimens were poorly preserved and never fully catalogued, and were ultimately scattered among collectors, museums and societies. Many of the avian specimens were simply skins or pelts with heads; the rest of the bird had been consumed or thrown away because, in order to be pickled in jars, they would have 'required more spirit [alcohol] than we could afford'.

Frankly, a budgerigar could have stowed itself aboard the *Endeavour* and chattered under the great botanist's nose, but he would still have paid more attention to a piece of antipodean bracken. The best a budgie might have hoped for was that Banks would eat it.

The Swedish naturalist Daniel Solander accompanied Banks on the *Endeavour*, and in 1782 he dropped dead of a brain haemorrhage in Banks's house while logging species discovered in the Pacific. The relationship between events may not have been entirely coincidental. That the task remained unfinished only added to the chaos of the Banks collection. In addition to material collected on the first Cook voyage, Banks, one of the eighteenth century's foremost men of science, was the recipient of many thousands of specimens sent by expeditioners and colonists around the world—first to his Mayfair home and, after 1777, to more commodious premises in Soho Square.

Reverend William Sheffield, as the keeper of the Ashmolean Museum at Oxford, oversaw a none-too-shabby collection that featured the death mask of Oliver Cromwell, the Kish tablet (inscribed with some of the earliest cuneiform/pictographic writing of the ancient Sumerian people) and the very lantern Guy Fawkes carried when he was arrested underneath the Houses of Parliament on that November night in 1605. Yet nothing could prepare him for his visit to Banks at the botanist's New Burlington Street home the year after he returned from Australia.

The good reverend could barely contain himself when recounting his excursion in a letter to fellow parson–naturalist and ornithologist Gilbert White on 2 December 1772:

> It would be absurd to attempt a particular description of what I saw there; it would be attempting to describe within the compass of a letter what can only be done in several folio volumes. His house is a perfect museum; every room contains inestimable treasure. I passed almost a whole day there in the utmost astonishment, could scarce credit my senses. Had I not been an eye-witness of this immense magazine of curiosities, I could not have thought it possible for him to have made a twentieth part of the collection.

Sheffield found the third room he entered the most breathtaking: 'This room contains an almost numberless collection of animals; quadrupeds, birds, fish, amphibia, reptiles, insects and vermes [worms], preserved in spirits, most of them new and nondescript. Here I was in most amazement and cannot attempt any particular description.'

Clearly, if there had been a budgerigar in there, it would have taken a long time to find it. But ornithologists believe that what was to become Australia's emblematic bird did not first reach Europe aboard the *Endeavour*. Their strongest argument in favour of this view is that budgies are generally absent from the coastal areas to which Cook and his company clung.

🐤 *Famous budgies*: DAIQUIRI

Victoria Legge-Bourke was at a garden party in London's Kensington Gardens in 1999 when she was swooped by a budgie. Better known as Tiggy, Miss Legge-Bourke was the nanny of Prince William and Prince Harry, and she provided love and security when their mother, Princess Diana, was killed in a Paris car crash.

The garden party was some years later and Tiggy was startled when the budgie plucked the straw from her daiquiri glass, took a long sip and promptly fell asleep. She adopted the budgie—naturally calling it Daiquiri—and it was visited several times by William and Harry at Tiggy's London flat.

By 1791, the British had pushed further inland and the very first definitive specimen of the natural green bird the world would come to know as the budgerigar was collected near present-day Parramatta.

That specimen—quite possibly first sent to Banks, who donated a large part of his collection of animals to the British Museum in 1792—was described in 1805 by George Shaw, the assistant keeper of the museum's natural history department, as a 'highly elegant species of Parrakeet', and named the Psittacus 'undulatus', because of the wave pattern on its wings. It featured in Shaw's opus

The Naturalist's Miscellany: Or, coloured figures of natural objects; drawn and described immediately from nature (1789–1813) accompanied by a life-sized hand-coloured copper engraving by Frederick Polydore Nodder.

The classification of the little green bird caused Shaw, a founding member of the Linnean Society of London, a lot less angst than some of the other specimens that landed on his desk during production of his 24-volume tome. Famously, Shaw was the first person to scientifically describe the platypus, based on a single skin and a sketch sent from the colony of New South Wales by the second governor, Captain John Hunter, in 1798. Shaw's incredulity shines through the *Miscellany* entry for the creature he identified as Platypus anatinus (flat-foot duck): 'Of all the Mammalia yet known it seems the most extraordinary in its conformation; exhibiting the perfect resemblance of the beak of a Duck engrafted on the head of a quadruped. So accurate is the similitude that, at first view, it naturally excites the idea of some deceptive preparation by artificial means.'

But it wasn't just weird and wonderful creatures from the Antipodes that were occupying the great scientific minds at this time. Over at the museum of the Zoological Society of London its curator and preserver, John Gould, was considering where to start on a gift of 450 birds and 80 mammals deposited by a young man named Charles Darwin on 4 January 1837.

Before the week was out, Gould excitedly reported to the Society that the samples included thirteen new finch species, none of which had previously been described. He noted that their 'principal peculiarity consisted in the bill presenting several distinct modifications of form'. The finches would subsequently play a major part in the development of Darwin's theory of evolution—to which Gould never publicly subscribed. By the time Darwin published his theory in 1859, Gould's reputation as an ornithologist was well-established and he did not care to court controversy.

This was a rare time in history when, thanks to the Victorian obsession with natural history, a gardener's son from the Dorset port of Lyme Regis could sidestep the strict taxonomy of the British class system. Gould's life had seemed charmed since he set up shop as a taxidermist in London in 1824 at the age of 21. Such was his skill that he became the first taxidermist by royal appointment after stuffing a thick-kneed bustard for King George IV. This was followed by commissions to stuff two deer, an ostrich and, most famously, a giraffe that had been a gift from the viceroy of Egypt.

 Famous budgies: SPARKIE WILLIAMS

The world's best-known budgie, Sparkie Williams, had a vocabulary of 583 words, including eight complete nursery rhymes. He came to attention after beating 2768 other entrants in a competition for talking birds organised by the BBC in 1958. He went on to front a birdseed campaign and make a record that sold 20,000 copies. On the B side of the Parlophone record Sparkie plays a gangster—'Sparkie the Fiddle'—and Australian actress Lorrae Desmond plays his moll.

Lorrae recalls she was at studios in London's famous Abbey Road to record her own LP when she was asked if she could put down a few words in a Brooklyn accent for the budgie's record. 'He was on the run from gaol and had just been shot and I had to say, "Oi Sparkie, they got ya real bad,"' Lorrae recalls with a laugh. The great bird wasn't in the studio at the time, but Lorrae later got to meet him when she was performing in his neighbourhood at the Newcastle Empire Theatre. 'He said, "Hello beautiful, give me a kiss."'

Sparkie toured the country with his owner Mattie Williams until he died, aged eight, in 1962. Acclaimed in the *Guinness Book of Records* as the world's most outstanding talking bird, Sparkie can still be seen today, stuffed, at the Great North Museum (Hancock) in his native Newcastle-upon-Tyne. In 2009 he provided inspiration for the Michael Nyman opera *Sparkie: Cage and Beyond.* When the opera was performed in Berlin, Sparkie was given leave to appear, and flew to Berlin accompanied by an archivist from the Natural History Society of Northumbria to ensure his safety.

With royal patronage and the support of men of science following his appointment to the Zoological Society, the ambitious Gould was a man quick to capitalise on opportunity. When a collection of previously undescribed birds came to the Society from northern India, he enlisted his wife Elizabeth's artistic skills to produce *A Century of Birds from the Himalaya Mountains.*

A second, more ambitious five-volume work, *The Birds of Europe,* saw Gould employ a young artist to assist Mrs Gould in the drawing of all her foregrounds. This artist's name was Edward Lear, and in 1846 he became famous as the author of *A Book of Nonsense,* a volume of limericks that went through three editions and helped popularise the form—and the genre of literary nonsense.

Even before the final instalment of Gould's book was published in 1837, he had the Great South Land in his sights, and had begun work on a new work to be titled *A Synopsis of Birds of Australia*. Much impetus was provided by Elizabeth Gould's brothers, Stephen and Charles Coxen, who had migrated to Australia in 1827 and 1834 respectively. Both Coxen brothers sent accounts and specimens of the 'strange and unusual' Australian wildlife to Gould, but Charles did so rather more expertly as he had also trained as a taxidermist.

In the months after his arrival in the colony Charles spent several months travelling the country between the Hunter and Namoi rivers collecting specimens. Many of these he sent to Gould, but he also provided some to the fledgling Australian Museum. In its very first catalogue, produced in 1837, the museum acknowledged that many of its fauna specimens, which included 36 mammal species, 317 birds, six fish and 211 insects, had been donated by the younger Coxen brother.

The birds of Australia were calling to John Gould with increasing urgency and in May 1838—accompanied by his wife, their eldest son, a teenaged nephew, two servants and the professional bird collector John Gilbert—he set sail for the Antipodes aboard the *Parsee*.

Chapter 2

John Roach, the scoundrel bird-stuffer

WHILE THE NINETEENTH century was certainly the golden age of natural history, and Sydney was soon to feed the English fascination with exotic flora and fauna, not all of those involved enjoyed a propitious path to the new colony. When the taxidermist John William Roach was barely twenty years old—just a year younger than Gould had been when he first came to the attention of King George IV—he found himself detained at his majesty's pleasure after being convicted in the Surrey Quarter Session on 27 May 1833 for stealing a coat.

A few days later he was one of 300 men and boys held aboard the *Hardy*, a reeking, lice-infested prison hulk that was moored in a foul Portsmouth backwater, with its portholes boarded up on the landward side to discourage any thoughts of escape. Roach and his cohort, which included 118 lifers convicted in counties across England, were awaiting transportation to Australia. As such, they were more fortunate than many other convicts of the period who spent entire sentences aboard the hulks.

On 26 June 1833 these prisoners arose blinking into the light from the depths of the *Hardy* and were transferred to the *Aurora*. They set sail for Sydney on 4 July, guarded by 27 members of the 21st Fusiliers under the command of the newly promoted Major Philip Delisle.

If their voyage was in any way remarkable, it was that no one died. However, there was no shortage of misery and privations, suffered particularly by the youngest convicts, William Banghurst and Thomas Westley, who were both just thirteen years old. By way of contrast, the only cabin-class passengers—Delisle's wife Sarah and his soon-to-be-married daughter Fanny—were troubled only by tedium and their delicate dispositions, the interminable creaking of the ship timbers and the veil of salt that became a second skin.

Roach, hardly more than a boy himself, was already in possession of more useful skills than many of those on board, who could scarcely scratch their names. His own initials, tattooed on his forearm, were noted as an identifying mark on his prison record.

In early November, as they were mustered on board the *Aurora* shortly after it arrived in Port Jackson, the prisoners' details— including age, education, religion, family, native place, trade, offence, sentence and prior convictions—were recorded before they were landed on 22 November and marched uphill to Hyde Park Barracks to be assigned for service.

John Roach at this time recorded his occupation as 'Bird Stuffer'. As such, he could hardly have found himself in the burgeoning colony at a better time.

It was that sensational oddity, the platypus, which had led to the founding of the Australian Museum after an urgent plea for some of its eggs to be sent to London. The recipient of this request, the newly installed Colonial Secretary in Sydney, Alexander Macleay, was a passionate naturalist who had brought his own extensive collection of moths and butterflies with him when he sailed to Sydney in 1826. He well understood the demand from naturalists at 'home' for new specimens, but his desire to promote science in his new land swiftly asserted itself. He had already been involved in founding the Australian Subscription Library and he now pushed hard for the establishment of a colonial museum.

Back in London, the Secretary for the Colonies, Lord Bathurst, responded positively: on 30 March 1827, he authorised Macleay's boss, Governor Darling, to spend £200 a year (roughly A$25,000 in today's money) on a 'Publick Museum', first called the Colonial or Sydney Museum. Intended for the display of 'rare and curious specimens of Natural History', it initially operated, as the *Sydney Gazette* reported on 31 August 1830, 'between the hours of 10 and three [for] . . . any respectable individual who may think fit to call'. It would be almost another 40 years until the Sydney Free Public Library, the precursor to the State Library of NSW, would join it as an institution.

By the time of John Roach's arrival in the colony, the fledgling Australian Museum had not long moved out of its first home—the shed and outhouse of what had been Australia's first Post Office, in Bent Street—and installed itself in rooms in the Legislative Council building in Macquarie Street. Here it came under the control of Edward Deas Thomson, a clerk who had in turn appointed his convict messenger, William Galvin, to look after the collection.

 Famous budgies: **BILLIE PEACH**

Australia's answer to Sparkie Williams was Billie (sometimes Billy) Peach. Billie had a vocabulary of close to 500 words and could sing 'God Save the King', 'Mary Had a Little Lamb' and 'Baa Baa

Black Sheep'. He was a radio star in the 1940s and also appeared on commercially released records, namely a 1939 home recording, *Billie Peach Part One & Two*. Billie also appears on film in the National Film and Sound Archive's collection in a 1947 documentary entitled *Time Off*.

When he was not making department store appearances in Sydney and Melbourne, Billie lived at home with his owner, Mrs Lydia Peach, and her husband in Darling Point. He was also a spokesbird for the RSPCA and a Red Cross Voluntary Aide during World War II.

He does not appear to be stuffed and on display anywhere. However, continuing the association between budgies and avant-garde contemporary music, his recordings were reinvented by musician, composer and multidisciplinary artist Heinz Riegler in a work titled *Billie Peach: Escape from Darling Point* in 2013.

Galvin, upon hearing of Roach's arrival in the colony, seems to have wasted no time in getting him on board as resident stuffer and collector of birds. By January 1834 Roach was ensconced in the museum, where he was charged with mounting the bird skins left behind by William Holmes—the colonial museum's hapless first custodian who had died on an expedition to Moreton Bay in Queensland 'collecting birds and other curiosities'. Holmes had shot a cockatoo with a single pellet, then accidentally shot himself while leaning down to retrieve the specimen.

So, less than a year after he had been shackled on the *Hardy*, Roach was not only allowed to carry a firearm and go hunting for birds, but he was also being paid 1 shilling and 9 pence a day (18 cents) in lieu of rations and clothing.

On 16 March 1836, Roach was attached to Major Thomas Mitchell's third expedition—just a day before it departed. It travelled to where the Lachlan, Murrumbidgee and Darling rivers joined the Murray, and led to the colonialists' discovery of the lush Victorian grasslands and sometime budgerigar playground that Mitchell called 'Australia Felix'.

The great explorer makes little mention of Roach in his extensive journals other than to note his inclusion in his party as 'Collector of birds'. Like all of the party, Roach was armed—in his case, ironically, with a 'fowling piece'—and issued with a 'suit of new clothing; consisting of grey trousers and a red woollen shirt, the latter article, when crossed by white braces, giving the men somewhat of a military appearance'.

But where Mitchell had next to nothing to say about Roach, his second-in-command, the dyspeptic surveyor Granville William Chetwynd Stapylton (grandson of a minor English aristocrat), was voluble in his dislike of the expedition's bird stuffer and whinged constantly about him across much of two states.

In entry after entry in his journal Stapylton railed against Roach, whom he described as an 'unhanged scoundrell'. A few examples suffice to illustrate the surveyor's poisonous malice:

The scoundrel Bird stuffer takes the greatest pains to conceal everything new from my sight. The collection for the Museum is already very extensive . . .

This damned bird-skinner has been spoiled in Sydney by Mr. Macleay, and is just the sort of free and easy vagabond with a flash shooting jacket on, that I feel especial pleasure in taking down a peg . . .

> *I've another Crow to pick with my friend the Bird-stuffer and it will be odd if I don't square accounts with him eventually.*

The expedition leader clearly had no time for his 2IC's petty gripes, as Stapylton indignantly noted on 23 September 1836: 'A complaint to [Mitchell] on the subject of the insolence of Roach he has the effrontery to call an exhibition of selfishness on my part and of a greater regard for my personal comfort than for the success of the expedition.'

Some days later Stapylton noted in his diary:

> *Grossly insulted again today by the Bird-stuffer. Ordered him to his tent. Swore he had as much right to the ground as I had, that I had no command over him, that he had as many pulls over me as I had over him and that he would make me remember my conduct towards him and a great deal more but I was in such a passion I cannot remember one half of it.*

Stapylton's time with Mitchell represented a high point in his career. After this he was sent south to work with the senior surveyor of Melbourne, where he was suspended from duty after several incidents of drunkenness. His complaints came abruptly to an end when he was killed by Aborigines outside his tent near the headwaters of the Logan River when assisting with the survey of the Moreton Bay area in 1840.

Meanwhile the fortunes of his antagonist continued to rise. Roach returned to the museum where, upon being granted his ticket-of-leave, he received an annual salary of £5 a month (about A$625 in today's dollars) as 'Collector and Preserver of Specimens', in addition to the £5 deposited in his name in the accounting for the Mitchell expedition. In August 1840 Roach left the museum and set up shop at 32 Hunter Street, no longer a crude bird-stuffer but now a free man, 'ornithologist and taxidermist', advertising no less in the endpapers of Joseph Fowles's pictorial book *Sydney in 1848.*

History does not tell us whether the first budgie John Roach stuffed was one he had collected during the Mitchell exhibition or whether he first mounted some that were legacies of earlier collecting by the unfortunate William Holmes. But the rascally bird-stuffer was to play a major role in popularising the little bird that had been a useful companion to Australia's First Peoples for many thousands of years.

Chapter 3

The Goulded cage

ONE LADY WHO had neither the time nor likely the inclination to peruse the windows of the bird-stuffers during her almost eighteen months in the colonies was Elizabeth Gould.

No well-bred young woman in all of Christendom had been subjected to as many dead birds as Eliza. Her union with John Gould saw her become his principal artist.

If there was one great joy to be had for a young mother suffering at her separation from three of her four surviving children—the youngest of whom was just six months old—to travel to Australia, it was that she finally got to sketch live birds . . . or at least when her husband was not in shooting range.

On the voyage to Australia in 1838 petrels, shearwaters and albatrosses brought down by musket or captured on hooks baited with lumps of fat had flapped at her feet as she sketched. Once the Goulds landed in Van Diemen's Land—as Tasmania was then called—the flutter of dead birds began in earnest, for Gould was a crack shot when the sea was not moving beneath him.

In the earliest surviving letter Eliza sent to her mother from

Hobart Town, she observed, not without pride, that her husband had 'already shown himself a great enemy to the feathered tribe, having shot a great many beautiful birds and robbed various others of their nests and eggs. Indeed John is so enthusiastic that one cannot be with him without catching some of his zeal in the cause, and I cannot regret our coming, though looking anxiously forward to our return . . .'

For Gould there was no conflict in being first to describe a new species for science by promptly killing it. He wrote of his first sighting of *Epthianura tricolor*, the colourful, nomadic crimson chat: 'As may be supposed, the sight of a bird of such beauty, which, moreover, was entirely new to me, excited so strong a desire to possess it that scarcely a moment elapsed before it was dead in my hand.'

Soon after landing in the colony, Gould began to see beyond birds. As he would later write in the introduction to *Mammals of Australia*: 'It was not, however, until I arrived in the country, and found myself surrounded by objects as strange as if I had been transported to another planet, that I conceived the idea of devoting a portion of my attention to the mammalian class of its extraordinary fauna.'

So, when the Goulds embarked on their homeward journey from Botts' Wharf, Darling Harbour, on 8 April 1840 aboard the *Kinnear*, it was with an extraordinary menagerie of mostly dead or soon-to-be-dead things. In addition to 800 specimens of birds, including the nests and eggs of 70 species, there were also 70 specimens of quadrupeds, and no fewer than 100 previously undescribed creatures preserved whole in alcohol for dissection. And, along with the ship's payload of wool bales, there were live animals and birds winched aboard, including several species of parrots. A solitary surviving pair of the smallest of these would make history.

Other passengers were apparently unaware of the treasures in their midst. James Johnston Macintyre, a Scottish businessman and traveller who kept a journal of his voyage on the *Kinnear*, noted the presence of the Goulds but not of their cargo, being rather more concerned with the health of a cow than cognisant of kangaroo.

Two months into the journey, Macintyre lamented the passing of the 'unhappy' bovine which had apparently been injured when being hoisted on board and had stopped giving milk a month before her death. 'The expense of getting her for the voyage and of laying in an ample supply of Derwent hay for her use were incurred by Captain Mallard and the circumstances of her dying so early makes us feel the want of milk more than if a cow had never been thought of for the voyage.'

While Macintyre confessed to doing very little at all while 'time glides away very imperceptibly at sea', the ornithologist and his wife kept very busy. When conditions were sufficiently calm Gould and his gun would be let down in a small boat to collect more seabird specimens while Eliza, who was still nursing her 'little convict' Franklin Tasman, born in May 1839, continued to sketch.

Eliza executed hundreds of drawings during her time away from the family home in Soho's Broad Street. She was responsible for producing 84 key lithographic plates for *The Birds of Australia* and the design of many others.

The drawing of *Melopsittacus undulatus*—the full binomial name taken from the Greek *melos*, meaning 'musical' and bestowed by Gould some 35 years after Shaw's initial classification—has particular poignancy, as these birds were drawn from live specimens that survived the journey to England longer than Eliza herself.

A year after the Goulds returned home, Eliza died at age 37, having contracted an infection after the birth of her eighth child.

Meanwhile the first budgies on British soil thrived. Gould later wrote that of the four little avian pioneers reared in captivity in the Hunter Valley by one of Eliza's brothers, two that had 'braved the severities of the passage to this country by way of Cape Horn in the midst of winter . . . are before me now in exuberant health'.

When budgies were formally and fulsomely introduced as 'warbling grass-parakeets' in *Parrots and Pigeons*, volume five of Gould's ornithological opus, it was to the highest society.

 Melopsittacus undulatus

Warbling grass-parakeet

Among the numerous members of the family of Parrots inhabiting Australia, this lovely little bird is pre-eminent both for beauty of plumage and elegance of form, which, together with its extreme cheerfulness of disposition and sprightliness of manner, render it an especial favourite with all who have had an opportunity of seeing it alive. This animated disposition is as Conspicuous in confinement as it is in its native wilds . . .

The beauty and interesting nature of this little bird naturally made me anxious to bring home living examples; I accordingly captured about twenty fully fledged birds, and kept them alive for some time; but the difficulties necessarily attendant upon travelling in a new country rendering it impracticable to afford them the attention they required, I regret to say the whole were lost. My brother-in-law, Mr. Charles Coxen, who resides on the Peel, having succeeded in rearing several, kindly presented me with four, two of which, as before mentioned, have reached England in perfect health.

As cage-birds they are as interesting as can possibly be imagined; for, independently of their highly ornamental appearance, they differ from all the other members of their family that I am acquainted with, in having a most animated and pleasing song; besides which, they are constantly billing, cooing, and feeding each other, and assuming every possible variety of graceful position. Their inward warbling song, which cannot be described, is unceasingly poured forth from morn to night, and is even continued throughout the night if they are placed in a room with lights, and where an animated conversation is carried on.

—John Gould, *The Birds of Australia*, vol. 5, *Parrots and Pigeons*, 1848

Queen Victoria, to whom *The Birds of Australia* was dedicated, headed the list of subscribers, which read like *Burke's Peerage*. No fewer than twelve crowned heads of Europe, sixteen dukes, 30 earls, five counts, 61 baronets and the Lord Bishop of Norwich signed up for the first 250 sets, published in 36 parts, which cost £115 (the equivalent of A$12,700 today).

It was in palaces that *Melopsittacus* perched.

Chapter 4

Birds in hand

SHREWD AS A shrike, John Gould recognised the value of the live birds. When he presented at scientific meetings the budgerigars went with him, twittering charmingly in their cage as he lectured. As word of these enchanting little creatures spread from men of science to their wives, Gould found himself a guest in the drawing rooms of London's most exclusive homes.

Caroline Owen, the wife of the brilliant naturalist Richard Owen, recorded in her diary for 27 March 1841 that Gould had 'brought his pretty singing New South Wales Parrots' along to a party held by Lord Northampton and attended by Prince Albert, husband of Queen Victoria.

The birds were coveted at the highest level of society, but Gould resisted pressure to part with them, fending off the keen entreaties of his patron, the thirteenth Earl of Derby, Edward Smith-Stanley, by saying they were 'especial pets' of Eliza's. In a letter to the younger of his brothers-in-law written several months before Eliza's death, Gould boasted of personally introducing the birds to Prince Albert at a meeting of the Zoological Society, adding 'I am any day expecting a Command from the Queen'.

The base from which 'The Bird Man' and his budgies sallied into high society was hardly salubrious. The Goulds' five-storey Georgian terrace in Soho was just around the corner from a workhouse, which was home to 530 people, and across the road from a brewery. Broad Street was also the epicentre of a major cholera outbreak in 1854 that killed 616 people. The eminent physician and founder of modern epidemiology, John Snow, sourced the outbreak to the Broad Street water pump, demonstrating it was the water supply and poor sanitation, not 'miasma' or bad air, behind cholera pandemics. In the course of his investigations he spoke to Gould and gleaned critical information.

 ### Budgies in a time of cholera

I inquired of many persons whether they had observed any change in the character of the water, about the time of the outbreak of cholera, and was answered in the negative. I afterwards, however, met with the following important information on this point. Mr. Gould, the eminent ornithologist, lives near the pump in Broad Street, and was in the habit of drinking the water. He was out of town at the commencement of the outbreak of cholera, but came home on Saturday morning, 2nd September, and sent for some of the water almost immediately, when he was much surprised to find that it had an offensive smell, although perfectly transparent and fresh from the pump. He did not drink any of it. Mr. Gould's assistant, Mr. Prince, had his attention drawn to the water, and perceived its offensive smell. A servant of Mr. Gould who drank the pump water daily, and drank a good deal of it on August 31st, was seized with cholera at an early hour on September 1st. She ultimately recovered.

—John Snow, *On the Mode of Communication of Cholera*, 1855

The feting of the erstwhile bird-stuffer of the Zoological Society who had risen to its vice-presidency astonished many of those acquainted with him because they found him coarse and uncouth. Edward Lear described his one-time employer's ascension: 'Gould—a man who, without any prospects or education, has by dint of a singularly active mind, good talents, and uncontrollable perseverance (not to say impudence) backed by no little good luck—risen in the world beyond belief.'

Lear, as one of 21 children, was obliged to begin earning a living as a natural history draughtsman from age sixteen after his stockbroker father was sent to debtors' prison. He illustrated the budgerigar in his very first book, the absolutely no-nonsense *Illustrations of the Family of Psittacidae, or Parrots*, published in 1832 when he was aged just twenty. While many of the drawings were from live birds of the Zoological Society or private collections, Lear's budgie was drawn from what was then the sole specimen in England: a stuffed bird held in the collection of the Linnean Society.

The book was critically acclaimed, and Lear's work was likened to that of the pre-eminent Bird Man of the Americas, Jean 'John' Audubon, but it was not a financial success. Lear, mindful of his father's example and the chill shame of penury, went back to working in the employ of others, including Gould, who purchased the unsold copies of Lear's books and incomplete drawings. 'I had rather be at the bottom of the River Thames than be one week in debt—be it ever so small,' Lear wrote emphatically.

Lear knew he had given Gould big ideas and was somewhat resentful. Many years later he wrote of him: 'He was one I never liked really, for in spite of a certain jollity and bonhommie [*sic*] he was a harsh and violent man . . . [A] persevering, hard working toiler in his own line,—but ever as unfeeling for those about him . . . He owed everything to his excellent wife,—and to myself, without whose help in drawing he had done nothing.'

Thanks to Gould's enthusiastic promotion, the demand for exotic Australian fauna continued to grow as private collectors competed for new specimens to join their menageries. Birds—alive or dead— were hot property and the burglars who broke into Gould's home in 1845 may well have been working to order. Among the treasures stolen and never seen again, in spite of Gould's advertised offer of a £20 reward, was a 'beautiful specimen of *Melopsittacus undulatus*'.

The great German bird collector and author Karl Russ credited one of his countrymen and contemporaries, the wildlife dealer Johann 'Charles' Jamrach, with selling the first pair of budgerigars in England—to a Dr Butler of Woolwich, for £27. (Jamrach's stepson, who worked with him in the business, later recalled the figure as £22.)

Charles Jamrach initially built up his wildlife dealership thanks to his father's contacts with sailors in his job as chief of Hamburg's river police. From the 1840s to the 1880s, Charles was the biggest dealer of exotic animals in the world.

Using agents in various European ports as well as all major British ports, he imported and exported animals for sale to zoos, private menageries and circus owners. Jamrach's Animal Emporium on the Ratcliffe Highway in the East End of London teemed with fabulous fauna including a Bengal tiger, which escaped its crate and seized a small boy before the proprietor famously wrestled the child from its jaws.

It is quite plausible that Jamrach also furnished the Earl of Derby with his first specimens of *Melopsittacus*. As well as maintaining several of his own collectors in the field, the earl was known to purchase material from more than 30 different agents and dealers. Irrespective of the origins of his stock, the aristocrat was indisputably the first person in England to breed them.

On 11 February 1848 his lordship gleefully wrote to Gould: 'I have pleasure to tell you that we have been overjoyed here by the fact of a Pair of the *Melopsittacus undulatus* breeding.' He reported that John Thompson, the superintendent of the earl's Knowsley Hall property and later superintendent of the Zoological Gardens in London, had observed that 'the hen never left the hole she had taken to but was regularly fed there by the male. He suspected she might be about hatching, but no broken shells or any other appearance could prove it and he very properly did not like to disturb her should it be so. We do not know anything more than that she certainly has hatched, for we can hear the young.'

Three weeks later the earl wrote again, this time to tell Gould 'both little Melopsittaci' had died. At least one of the chicks was stuffed and preserved in the earl's museum at Knowsley and later bequeathed to the people of Liverpool as part of a collection of 20,000 birds and animals, including the dodo, now housed in the city's World Museum.

Chapter 5

Further adventures of the bird-stuffer

WHEN OUR OLD mate Roach moved from his first modest shopfront to hang his shingle in Sydney's Lower Hunter Street, the area was already a hub for taxidermy and trade of Antipodean fauna. No fewer than six businesses devoted to selling all manner of creatures, both alive and dead, operated there between the 1840s and 1850s. Sydney's first circus, a marquee called the Australian Olympic Theatre that featured a troupe of Mauritian acrobats and gymnasts, briefly added to the exotic mix of the Hunter Street scene.

Cheekily, Roach set himself up in opposition to his old employer: when his new emporium of curiosities opened in August 1843, he styled it as a 'museum'.

 J. W. Roach

Taxidermist, collector and preserver of specimens of natural history BEGS leave to announce he has fitted up part of his spacious premises, No. 4, Hunter-street, as a MUSEUM, containing specimens in every branch of natural history, comparative anatomy, and curiosities of all descriptions to which additions will be constantly made . . .

The collection already contains, amongst others, some hundred specimens of the birds and animals of Australia, set up, and arranged in their most natural and pleasing forms.

The want of such a repository has long been felt by the scientific and curious, and J. W. R. trusts that his collection, to which he respectfully invites the attention of the public, will meet the approbation of all who are engaged in the interesting and pleasing study of the most beautiful and perfect works of the Creator.

Open to the public every Monday and Friday, from 11 o'clock, a.m., until 6 o'clock, p.m. Admission, one shilling. Children under Twelve years of age, sixpence. Cards of admission to be had on the premises.

J. W. R. has also an AVIARY, containing living specimens of some of the rarest and most beautiful birds. Collections supplied to order. Specimens set up and arranged in the best style, on moderate terms.

—The Australasian Chronicle, 26 August 1843

The exotic fauna of New Holland was the subject of much fascination, not only among men of science but also in the fashionable drawing rooms of London. Long before the first live budgerigars were landed upon English shores, whole flocks of Australian birds—including the first emu, collected in 1788 from what is now Redfern—had been skinned, swabbed with arsenical soap and returned to the mother country where, in grim dioramas, they perched in perpetuity in upper-class homes.

Nor were the colonists themselves immune to fashionable flutter. As Dr George Bennett, who became the first secretary of the Australian Museum in 1836, observed of parrots on his second arrival in the colony in 1832: 'No one can walk the streets of Sydney or any villages of the colony, or enter an inn or dwelling-house, without seeing this class of birds hung about in cages, and having his ears assailed by the screeching, babbling and whistling noises.'

The natural history trade was a highly competitive business and Sydneysiders then, as now, were ever hungry for the new. Before Gould introduced the wider world to *Melopsittacus undulatus* in volume five of his opus *The Birds of Australia*, Roach was not only enthusiastically buying and selling them, but also became the first to identify them as 'budgerigors' [*sic*] in any published form.

To settlers and residents in the interior

J. W. ROACH begs to intimate to the residents of the interior that he is a purchaser of LIVE BIRDS, in any quantity, as follows:—

Plyctolophus Eos, or Kilaw of the Aborigines; the Rose Parrot of the Colonists.

Platycerus Novæ Hollandæ, or Cook's Crested Parrot of the Colonists.

Nanodus Undulatus, or Budgerigor of the Aborigines; the Shell Parrot of the Colonists.

Menura Superba, or Mountain Pheasant or Lyre Tail of the Colonists.

Alectura Lathami, or Bush Turkey of the Colonists.

Also, the Eggs of all the above-named Birds.

J. W. R. begs likewise to inform Captains of vessels and Gentlemen about to leave these shores for Europe, India, or the neighbouring colonies, that he has always on hand Black Swans, Emus,

Kangaroos, and many other varieties of Birds and Animals, for sale at his Repository, Hunter-street, Sydney.

—*The Sydney Morning Herald*, 27 January 1845

When the wealthy young tourist and artist Eugène Delessert strolled into Roach's shop one idle afternoon in 1845, he was so charmed by a budgie that he wrote of the encounter in his book *Souvenirs d'un Voyage à Sydney (Nouvelle-Hollande)* published upon his return to Paris in 1847:

> *A man named Roach, who has a great reputation as a bird stuffer, and who receives numerous orders from Europe, keeps a curio-shop in Hunter Street, worthy of attracting the attention of strangers, particularly those who, making only a short stay in Sydney, haven't the time to go into the surrounding bushland. At Roach's one can have the pleasure of seeing all at once a selection of the animals found in New South Wales ... But the parrakeet [sic] which is the tiniest, rarest, and consequently the favourite, is the one called budgerry. It is the size of a canary, clear leaf-green in colour and striped with black on the back. Nothing is more amusing than to hear it chatter and ask for a piece of bread. It is a bird which can be taught without too much trouble.*

Roach's business appears to have enjoyed an edge over competitors, not always entirely legitimately.

In September 1846, Roach once again found himself on the wrong side of the law after falsely representing himself as the curator of the Australian Museum and removing the foetus of a dugong or 'sea

Budgerigar

pig' from a boat just arrived from Moreton Bay. The legitimate and furious curator, William Sheridan Wall, found the valuable specimen in Roach's shop. There was no love lost between the men—they had briefly shared the title of Collector and Preserver of Specimens at the museum before Roach left the institution's employ—and Wall alerted the law. Roach appears to have talked his way out of prosecution, but the following year was at the forefront of another even more bizarre fraud.

In October 1847, newspapers carried a scathing dissection of Roach's latest star exhibit:

 ## The bunyip unmasked

The head of this 'mysterious monster', now exhibiting in the window of Roach, the taxidermist, in Hunter-street, and which, for a considerable period, has excited the wonder of our unsuspecting townfolks, and gullible country cousins, has indubitably been proven to be the malformed cranium of a foal. This *lusus naturae* was foaled at Windsor, to our certain knowledge, on the farm of a gentleman, to whom reference will be given on application at this office.

The mother, a valuable animal, was compelled to be shot, owing to the impossibility of giving birth to this abortion; and the head of the latter having been separated from the trunk, was carried away by a quack horse doctor in attendance, and it is supposed subsequently handed over to Mr. Day, connected with the establishment of Youngman and Co., the druggists, next door neighbours to the credulous Mr Roach!

Such is the true history of this MANCHAUSENIC WONDER [sic], and if any doubt the accuracy of this statement, we shall be

32

happy to afford them every information necessary to obtain a view of the skeleton, which is buried adjacent to the spot where the mare was destroyed in her unnatural labour.

—*Bell's Life in Sydney and Sporting Reviewer*, 23 October 1847

The wily Roach was, it seems, the archetype of the bunyip, in its colonial sense of an imposter or pretender.

By 1849 the 'rascally bird-stuffer' had decided there were easier ways to make a living and at age 36 joined the rush to California for gold, together with fellow ex-convict and Hunter Street natural history dealer James Palmer.

Before sailing, Roach instructed Sydney auctioneer Charles Newton to sell the contents of his business and home and the 'whole of his well-selected stock of birds, quadrupeds and curiosities'. The advertised sale included the contents of 'two large aviaries containing a great variety of parrots, pigeons, paroquets, budgeree ghas' as well as an assortment of birds both skinned and stuffed and 'about 150 gross animals and fishes' eyes assorted'. The items on the list showed how colonists also blithely traded in Indigenous human life, including 'a number of skulls of the natives of various islands— these will be found to be worthy the attention of phrenologists. A lot of clubs and war instruments'.

Hunter Street remained a hub for the natural history trade well into the 1860s, as others quickly occupied the space left by the gold fevered. Writing of her eighteen months spent living in the colony, Emma Macpherson, whose husband Allan was a member of the Legislative Assembly and owner of properties in New South Wales and Queensland, was keen to give a 'notion of everyday life in the colonies, as it would appear from a lady's point of view'—untold, as it were, by works of far higher pretensions by the statesman, the man of science or the emigrant. Helpfully, this included a critique of Sydney shopping and the admonition to wear stout shoes.

'The most important of these thoroughfares is undoubtedly George Street—the Regent Street end of Sydney—in which most are to be found,' she wrote in 1857:

Intersecting this at all sorts of angles are other streets, generally of less importance, but some of them boasting of many good shops, as in the case with King Street and Hunter Street. The last, indeed (Hunter Street), contained two which had more interest for me than any others in the town. They belonged to birdstuffers, and their windows were always full of many curious and beautiful specimens of the feathered natives of Australia, both alive and dead: parrots and parroquets of many and various hues, the well-known white cockatoo, with its lemon-coloured crest, and the far rarer and larger black one, with bright scarlet top-knot and tail-feathers, a most magnificent bird, specimens of which are most difficult to be obtained, from its wild shy habits.

These and many other denizens of the bush, down to the well-known boodjerigah, or shell parrot, or love bird, as it is more commonly called, with its beautiful plumage of soft vivid green, and the still smaller and scarcely less beautiful diamond bird [diamond firetail] with its speckled wings and golden breast, made these shops points of great attraction to me, and I plead guilty to many idle moments passed in gazing in at their windows, and admiring and even coveting their contents.

Postscript: The bird has flown

By 1858, John Roach had returned from America. He was listed as a steward and proprietor of the German Club in Wynyard Street in the *Sydney Directory* in 1858 and in 1861.

The last known record is in the 1864 *NSW Police Gazette* where he was reported having escaped from the custody of Michael Kelly of Mudgee Police after being charged on warrant with assaulting Joseph Gunner of Biraganbil, near Gulgong.

He was described as being between 50 and 60 years of age, 5 feet 8 inches (172 centimetres) tall, with a stout build, round shoulders, sallow complexion, black hair mixed with grey, short black whiskers and beard of about a fortnight's growth. He was dressed in moleskin trousers, striped grey Guernsey shirt and a hat covered in white calico. He was thought to have made for Dubbo, Maitland or Wellington.

Chapter 6

Budgerigar Dreaming

EARLY SETTLERS WERE quick to dub brolgas the 'Native Companion', but the stately cranes were mere camp followers when compared with budgerigars. Long associated in Aboriginal lore with the getting of wisdom and rites of initiation, the little green and yellow bird helped Indigenous people read their environment, heralding rain and signposting food and water.

While well known to the riverine people, budgerigars held greater significance for the nomadic desert groups who observed their seasonal movement across the harsh heart of the continent.

Captain Charles Sturt's unshakeable belief that a great inland sea lay at Australia's heart was deduced, quite erroneously as it transpired, from budgies. 'I had seen the *Psytacus Novæ Hollandiæ* [Psittacus] and the *Shell Parroquet* following the line of the shore of St. Vincent Gulf like flights of starlings in England,' Sturt wrote. 'They all came from the north and followed in the same direction . . . a reasonable inference may be drawn from the regular and systematic migration of the feathered races.'

Three gruelling expeditions yielded no sea, but as sometime explorer Alexander Magarey observed in an 1895 paper, 'as thoroughly reliable guides to water the birds have no rivals. Some varieties should be named the bushman's water finders'.

As well as being bellwethers, budgerigars made good eating, with both the eggs and baby birds considered as delicacies across tribal groups whose territory they shared. Men used branches, killing sticks or boomerangs to bring down birds in flight. They also used nets suspended on poles across waterways to catch a variety of birds, as Thomas Mitchell observed when travelling along the Darling and Murray rivers with his surly surveyor and smart-alec bird-stuffer in tow. Women made the nets from reeds and bark, soaking them in water and pounding them so the fibres were pliable enough to plait.

Walter E. Roth, a colonial administrator, doctor and anthropologist who served as first Northern Protector of Aborigines and later as Chief Protector, observed the ingenious 'net and alley' approach to catching *Melopsittacus* among the Pitta Pitta people in central-west Queensland. In his ethnological opus, published in 1897, he wrote:

Stretching from some water-hole two long divergent palisades are built; these are made with thick bushes, saplings and twigs about eight or ten feet high, and forty and fifty yards long. The space in the narrower portion of the alley is cleared of trees, those in the diverging portion being left untouched.

In the very early morning a number of men sneak up toward the trees . . . and with loud shouting and every kind of noise will suddenly commence throwing sticks and boomerangs into them. The birds being thus driven from their roosts, by what they think to be hawks, fly low and in a direction opposite to whence the noise proceeds, but not being able to penetrate the bushes forming the palisade, make straight for the waterhole, where they are intercepted in scores by a fine-meshed net held up by two men standing just in front of it.

One of the most detailed accounts of Aboriginal hunting and gathering practices was provided by anthropologist Donald Thomson, who spent many months living among the Pintupi people—then known as the Bindibu—of the Western Desert of central Australia in the 1950s and 1960s. These were the last people living a traditional nomadic lifestyle in some of the harshest country on Earth, with the Pintupi Nine famously coming in from the desert in 1984.

Thomson, who briefly worked as a cadet journalist, was one of the most important chroniclers of Aboriginal life in central Australia and across the Top End. He also took some 4000 black-and-white glass-plate photographs, including a photo of a group of men and their bark canoes that became the inspiration for the award-winning Rolf de Heer film *Ten Canoes*.

Fortunately, despite snarks from some academic colleagues about being lowbrow, Thomson believed it was important to share the story of Indigenous people with everyday Australians in accessible articles, and penned the following for the *Daily Telegraph*.

 ## Budgie banquet

The cooking or preparation of these kilkindjarri (budgerigars) was an eye-opener to the white man, accustomed to the waste involved in the preparation of poultry and larger game. For the Bindibu [Pintupi] can afford to waste nothing. The birds were thrown onto the ashes of the small cooking fires and the feathers singed off, after which they were brushed and this process was repeated until the bodies of the birds were free of feathers.

The whole bodies were eaten, including the bones, even the skull and brains. The bill was set aside with certain other parts that were to receive special attention later. The horny layer, yellow or orange in colour, which formed the outer covering of the bill, was then stripped off and discarded, the bill itself being eaten with the whole of the skeleton of the bird, except the carina or breast-bone. Even the viscera were picked over and most of these eaten. I cannot say that the people always eat almost the entire skeleton of these small parrots, but at this time they were hungry for kuk'a— for animal protein.

—Donald Thomson, *The Daily Telegraph*, 1958

Budgerigars famously feature in the Jukurrpa—or, as western anthropologists coined it, The Dreaming—depicted in the art of the Warlpiri peoples whose country is centred in the Northern Territory's Tanami Desert. Here, the budgerigar is known as *ngatijirri* and it is a recurring motif in the art of its custodians, expressed by both footprints of the birds on the ground and cross-section views of the birds in flight.

Famous leaders of the desert art movement, including Anmatyerre men Billy Stockman Tjapaltjarri and Kaapa Tjampitjinpa, have painted the Budgerigar Dreaming, as has Johnny Possum Tjapaltjarri, half-brother to the internationally renowned Clifford Possum. Women artists who share this dreaming include Peggy Rockman Napaljarri and Myra Nungarrayi Herbert (also known as Myra Nungarrayi Patrick), who recalls listening for the chirping sounds of birds in nests to hunt budgerigars around Tanami as a young woman.

While the ngatijirri Jukurrpa can be mapped to specific locations—most frequently waterholes in traditional Warlpiri lands—budgerigars feature in the storytelling of other communities and language groups across the areas of the birds' distribution. The Kalkadoon or Kalkutungu people of Queensland's Mount Isa region know the indigenous bird as *tjinparra*. To the Eastern Arrernte people living on the sunrise side of Alice Springs it is *atetherre*, and the Noongar peoples of south-west Western Australia call it *dingleyerung*. The Ngameni people, whose land stretched from far north South Australia into south-west Queensland, called the budgie *katatara*, and initiated men wore long girdles of hair string with its feathers interwoven.

Despite the diverse roles budgerigars play in the stories of Indigenous communities across much of Australia, all have common origin in the time when the land, all its people, animals and vegetation were created by the Spirits.

Versions of the Dreaming stories recorded by early white settlers reflected the belief that these were childish fairytales rather than part of an instructional narrative and complex totemic blueprint. The Budgerigar Dreaming is detailed right down to the individual deposit of yellow ochre created by the falling feathers of the ancestor spirit. But stories presented in the popular press as Aboriginal myths in the early part of the twentieth century were frequently highly romanticised, if not pure hogwash, like the story below.

Romantic legend

The Queensland aborigines, before they passed into extinction—there are very few of them left today—treasured a legend of the budgerigar, a simple story instinct with the poetry of the Greeks.

It was the story of a boy and his lubra, Budgeri and Gar, whose love was within the bonds of totem and therefore forbidden by the tribe. Rebels against convention, Budgeri and Gar ran away from the campfire together, pursued by the old men with the nulla-nullas of revenge, slipping unseen through the bushland, until a great river barred their way.

In the nick of time their plight attracted the attention of Biamee, the Boomerang-Thrower, Lord of the Thunder, Zeus of the Southern Skies, who changed them into two little bundles of bright feathers, to fly away from the tribe forever, alone together.

—*The Advertiser* (Adelaide), 23 July 1934

The ancestral budgerigars are closely affiliated with water and their appearance is more often a pragmatic clue to map the vital resource or ceremonial site rather than following the arc of European storytelling. Young desert men are taught the ancient wisdom of the Budgerigar Dreaming path, learning key locations through songlines and sand drawings during initiation ceremonies. Being constant chatterers, the birds also often bear witness to key events of Aboriginal lore.

In 1966 anthropologist Mervyn Meggitt published a translation of the long tale of two Warlpiri (or Walbiri, as it was then spelled) ancestral spirits known as the Mamandabari as they criss-crossed Country, passing through totemic sites, sometimes travelling through the air, sometimes underground, performing ceremonies, singing of their encounters and many trials. Eventually the heroes, severely worn by their journey, are brought down by dingo men. At the very end waits the budgie.

Budgerigar

 ## Ngatijirri knows

As the dingoes disembowel the Mamandabari, they worry the corpses so vigorously that the men's hearts are tossed high into the air to fall nearby where they are transformed into two big stones lying in a rock hole. Sated at last the dingoes build a great fire over the few remains of the Mamandabari's corpses in order to conceal their dreadful deed, then lope away silently.

Meanwhile, a little budgerigar attracted by the commotion, has flown to the scene. Sitting silent and unnoticed in a tree, it watches the slaughter. After the dingoes depart the bird loudly mourns the death of the two heroes. Later it travels about the country and tells other people what has happened. That is why the story is known today.

—Mervyn Meggitt, 'Gadjari among the Walbiri Aborigines of Central Australia', *Oceania*, 1966

Celebrated in the paintings, stories, songs and ceremonies, the budgerigar ancestor naturally had its rightful place in the heavens in Aboriginal astronomy. People across the land knew the sky and attached names and stories to many of the fixed stars and star groups.

The Wiradjuri of central New South Wales, for example, know a smallish star in the constellation, formerly known as Argo Nevis, as Gidyirrigaa. *Gidyirrigaa* is the Wiradjuri word for budgerigar, and to the neighbouring Gamilaraay (Kamilaroi) people it is the same but for one letter: *gidjirrigaa*.

While the precise etymology of the word *budgerigar* has not been established, there is no question it was one of around 400 words borrowed from Aboriginal language in the first 100 years of colonisation. According to one school of thought, it is a loan blend from the New South Wales pidgin English word *budgeree* or *budyari*—first recorded in the 1790s as meaning 'good', 'well' or 'excellent' in the traditional language of the Dharug people living in the Sydney basin—then compounded with *gar*, which has been variously taken to mean 'little', 'bird' and 'food'. So, in this version, the word has generally been interpreted to mean that the bird made good eating.

But this seems to ignore the fact that the budgie is not indigenous to the Sydney region and was rarely seen in coastal areas except in time of drought.

What is indisputable is that early naturalists and explorers wrestled hugely with the spelling of Aboriginal words, with variant early spellings of *budgerigar* including betcherrygah, budgerri-gang, betshiregah, budgerigor, bidgerigung, budgeree gaan and betshirygah.

Encountering his first live specimens of the bird that he would name *Melopsittacus undulatus*, John Gould noted the 'Natives of Liverpool Plains' (part of the great nation of Gamilaraay people) called it *betcherrygah*. The colonists at this point knew the bird as the canary parrot, and Gould styled it the Warbling Grass-Parrakeet.

In the introduction to *The Birds of Australia*, Gould wrote that 'the beautiful little warbling Grass-Parrakeet (*Melopsittacus undulatus*), which, prior to 1838, was so rare in the southern parts of Australia that only a single example had been sent to Europe, arrived in that year in such countless multitudes on the Liverpool plains, that I could have procured any number of specimens, and more than once their delicate bodies formed an excellent article of food for myself and party.'

Gould was always keen to learn the Aboriginal names for the specimens he collected. This was not so much because he desired Australian birds to be known to the world by their Indigenous names, but because Aboriginal people could help collect specimens if they knew what he wanted.

Explorers and naturalists relied heavily on Aboriginal people to inform and guide their expeditions. Gould included two Aboriginal men, Natty and Jemmy, in his party when he explored central and northern New South Wales. He considered them invaluable. 'I find the natives useful in assisting, being scarcely ever without a tribe or portion of a tribe with me when in their neighbourhood; they are nearly all excellent and dead shots, and are excessively fond of shooting,' he wrote in 1839. 'I frequently give into their hands my best guns, and never find them in the slightest degree disposed to take advantage.'

He also urged his collectors to gather intelligence from local people. 'I am particularly anxious that you should obtain on the east coast and in New South Wales, even about Sydney, as many of the aboriginal names of Mammals and Birds as you can, particularly the origin of the word Kangaroo,' Gould wrote to his principal collector, John Gilbert, in 1844.

This stratagem did not always end well. Three of Gould's collectors, including Gilbert, were killed for their trouble—but not without provocation.

Gilbert was fatally wounded on 28 June 1845 when a hail of spears pierced the night camp of Ludwig Leichhardt's first expedition. It was a revenge attack by Cape York Peninsula Aborigines, after several of their women were raped or molested by members of the explorer's party.

The following month Johnston Drummond, son of botanist James Drummond, had two glass-tipped spears driven through him as he slept while on a specimen-collecting journey at Moore River in Western Australia. His assailant Kabinger—a Noongar man who had previously guided both Drummond and Gould—then reclaimed his wife, whom Drummond had taken as his own sleeping partner. Kabinger was later tracked down and killed by Drummond's younger brother John Nicol, who was the Western Australian colony's first Inspector of Native Police.

The third of Gould's collectors to die was Frederick Strange. He was killed along with three other Europeans by Dharambal men while collecting shells on the beach on Middle Percy Island, off the central Queensland coast. The motive was unclear but may have involved some unwitting transgression by Strange who, according to the party's surviving Aboriginal interpreter, was suddenly speared in the leg—usually a deliberate act as punishment or warning. Strange's response was to shoot dead his assailant. His own death and those of his companions swiftly followed.

The Aborigines, recognising the white man's mania for wildlife specimens, also used their superior knowledge and skills to independently collect birds and animals in exchange for goods or money, as the entitled little Frenchman Eugène Delessert noted in his journal. 'With great skill they can find parrots' nests, whose eggs they eat. As for these birds' little ones, they raise them to sell them in the towns, and this it seems is the only industry in which they indulge. It often happens that an Aborigine will cover twenty leagues to sell a parakeet for some feeble sum of money.'

But Delessert, who prided himself on his travel reportage being without 'exaggeration or prejudice', was no mere observer. He, too, was driven by the need to collect and had a family reputation to uphold as the son of Adolphe Delessert—a naturalist who travelled extensively in India and Southeast Asia and had the Wayanad laughingthrush (*Garrulax delesserti*) of southern India named after him. His great-uncle, Baron Benjamin Delessert, was even more distinguished, having established one of the richest herbarium and conchological collections in Europe with more than 250,000 specimens of plants and 150,000 shells. The baron's early education in botany included instruction and specimens provided by another keen amateur biologist and one of the fathers of the Enlightenment, the philosopher Jean-Jacques Rousseau. He was also a brilliant businessman and genuine philanthropist who founded and funded the first soup kitchens in Paris.

Young Eugène did not do well by comparison. He was deeply judgmental of Aboriginal people and peevishly wrote of their reluctance to give him what he wanted. 'It is not without trouble and sacrifice that I managed to obtain most of the weapons in use among the savages of the interior of New-Holland; they do not wish to part with them either for money or in exchange for anything,' he complained.

Much of what Delessert collected in travels in the Pacific, China, Japan, New Zealand and Australia subsequently ended up in the Muséum d'Histoire Naturelle in his home town of Le Havre, northern France. And, shamefully, Indigenous peoples were themselves souvenired, their very bones counted among the curiosities in the colony's burgeoning natural history trade.

Chapter 7

How ornithologists tell it

AS FINDS GO, the fossilised foot bone of a budgie isn't much: just 10.5 millimetres, to be exact. But this tiny tarsometatarsus represents the oldest evidence of *Melopsittacus undulatus* ever found, and it created quite a flutter among ornithologists when it was revealed by Dr Walter Boles in the peer-reviewed scientific journal *Emu* in 1998.

In all, four budgie bones have been recovered from the famous fossil ground Riversleigh, in north-west Queensland. The bones were found at Rackham's Roost, one of more than 200 sites within the World Heritage-listed area, and described by the great naturalist Sir David Attenborough as one of the four most important fossil deposits in the world.

Here in a cave above the Gregory River, some 2 million years ago, these budgies were snatched from their sleep and met a grisly end as ghost bat fodder. 'The gleaning bats, which are much bigger than budgies, would fly out at night and grab birds that were roosting, take them back to the cave and chomp on them,' Boles, a senior fellow at the Australian Museum, says. 'They tended to eat them straight down the middle and drop the wing and leg bones.'

Based on the most recent radiometric dating of the layer above the fossil bone layer of the ancient roost, the bones are estimated to be at least 1.1 to 2.8 million years old. Together with the beak of a white cockatoo dating from the earlier Miocene era (23 to 5.3 million years ago), they comprise the earliest record of parrots in Australia.

Unlike most other fossil sites that are aquatic-based, the relatively dry, protected environment of the cave helped preserve the remains of the budgies, and of other small birds and animals that made up the ghost bats' nightly repast. 'In old lake beds and the like you don't get little birds, so unless they find another ghost bat accumulation the chances of finding older small parrots are limited,' Boles says. 'A lot of it is just vagaries of the fossil record.'

That is not to say budgies haven't been around a whole lot longer than 2 million years. Budgies, like all birds, evolved from theropod dinosaurs of the two-legged kind, like the tyrannosaurs and velociraptors. The skeleton of the first definitive bird archaeopteryx, regarded as a transitional fossil between dinosaurs and modern birds, was unearthed in Germany in 1861.

After dinosaurs became extinct 66 million years ago, bird diversity exploded. The two earliest branches of the bird family tree to split apart represent birds that are found only in Australia and New Zealand, including the earliest of all parrots.

In August 2019 scientists revealed they had discovered a major piece in the parrot paleontological puzzle in the form of a one-metre tall specimen. The fossilised bone fragments of a bird initially thought to be an eagle were found near St Bathans on New Zealand's South Island—one of the richest deposits of creatures from the early Miocene period. When, after sitting on a shelf for more than a decade, the fragments were finally scientifically described, the evidence screamed megaparrot, or 'Squawkzilla' as it was dubbed in the press. The researchers responsible for this breakthrough gave the bird the name *Heracles inexpectatus* . . . because no one expects a 7-kilo parrot.

Scientists estimating the divergence times among various parrot groups believe the split that gave rise to the modern budgie occurred some 20 to 22 million years ago. 'At that time, Australia would have been covered with rainforest and it is highly likely that the early budgie ancestor was a rainforest bird. When Australia began to dry out in the late Miocene, the rainforests retreated and were replaced by woodlands, with grasslands starting to appear and spread,' Walter Boles explains.

'There is no indication of how diverse the budgerigar lineage was at this point, but one part of it adapted to the new food source of grass seed. This likely started around 5 to 6 million years ago, although there is no direct evidence to support this statement. While many birds failed to adapt to the changing environment, the early budgerigar became very successful in the evolving setting and went on to spread across the dry sections of the continent, becoming the bird we know today.'

Budgies, like other monotypic parrots—the ground parrot and enigmatic night parrot—are the only species in their genus. For many years it was thought they were akin to the rosella, but recent DNA analysis shows they are most closely related to lorikeets.

'It is an interesting little bird because it fooled us for so long by looking and behaving like something it was not,' Boles says admiringly. 'It has had so much attention given to it by breeders over time we know an awful lot about their genetics now. It is the number one cage bird in the world and yet it carries on as the quintessential Australian bird in the wild.'

The budgie is supremely well suited to life in the arid and semi-arid interior of its homeland. Highly nomadic, it follows rain and feeds predominantly on seasonal seeding grasses. In winter months it is to be found in Queensland at the base of the Gulf of Carpentaria, migrating back to the centre of Australia to breed in spring and summer. The booms and busts of budgie populations are based on the cycles of food supply that follow the El Niño effect. Breeding among the red gums and river gums along river courses, populations explode into millions in the good times and perish in the bad.

Budgigeddon

In January 1932 a heatwave gripped central Australia, with temperatures averaging from 46.7 to 52 degrees Celsius for sixteen successive days. At Paratoo, about two and a half hours north-east of Adelaide, witnesses reported how budgies, escaping the ferocious heat, darkened the sky from 5 a.m. to 8 a.m. as they passed overhead at an estimated rate of 1–10 million every 10 minutes.

Millions more didn't make it, dying of heat and thirst. At Kokatha Station, one of the westernmost homesteads in South Australia, the scene was devastating. Station owner Reg Wilson, speaking to Adelaide's *Advertiser* in May 1932, reported removing 5 tons of parrots from a single dam.

'We made a net with wire netting and dragged it from one side of the dam to the other, and then extracted the birds, as a fisherman does his fish. We filled a petrol tin and weighed it, and then by counting the tins were able to arrive at the weight.

'From one of the tanks we took out 30,000 dead parrots. The birds would settle on the water for a drink; others would follow and push them under the water; and this went on until dead bodies were many inches thick.'

In the 1940s celebrated French ophthalmologist André Rochon-Duvigneaud neatly defined a bird as 'a wing guided by the eye'. This is certainly true of budgies, which have such incredible long-range vision they can see rain hundreds of kilometres ahead and navigate to the area with that knowledge. They usually migrate at night and, harnessing tailwinds, can cover 600 to 800 kilometres in a day.

Melopsittacus is hardwired to water. Even after more than 170 years of captive breeding, caged birds will respond to rain and even running taps with excited chirping. Water means food, which in turn signals breeding, and birds respond with noisy urgency as hormones, dormant during drought, are stimulated. Some owners spray birds with water and even shower with them to trigger breeding.

Experimentally, budgies have been found to survive up to 38 days without water, giving rise to one of the more common misconceptions about the bird. 'A budgie won't survive without water for more than a few days unless it has some moisture in its food,' avian health expert Dr Rob Marshall explains. 'In dry weather they sustain that capacity by feeding on grasses beneath the mulga tree. There is a micro-environment underneath the mulga tree, which are the most populous trees in Australia and dominate the arid zones. Budgerigars gain their water underneath the tree in the humidity zone where you get little grasses producing moist green seeds.'

The budgie has an intelligence-based survival mechanism. 'It has to change its thought patterns every day,' Marshall says. 'The key thing that separates budgerigars from other parrots is their adaptability. It is the most adaptable parrot in the world. The species' ability to survive so long and in such harsh conditions is based on their intellectual capacity to respond, to think and know what is going on. Really, they are just amazing, very, very clever birds.'

'Parrots in general have the largest brain relative to the size of their body of any bird that exists today,' Dr Andrew Iwaniuk agrees. The Canadian biologist undertook much of the research for his PhD in Australia, where he studied the evolution of the brain in birds, and he says there is much still to be understood about budgerigars, particularly in their social behaviour and cognition.

'In contrast to lots of other Australian parrots of the same size, they are far more social,' Iwaniuk says. 'Turks [turquoise parrots], scarlet-chested, blue wings or any of the smaller grass parakeets tend to be very territorial and are bastards to keep in aviaries because they want to defend the whole cage environment as their own.

'With budgies, when resources are very low and concentrated in specific spots, then you will have these huge flocks and the flocks will undergo fission-fusion, and when there are lots of resources—water, grass seeds, areas to forage—they split up into smaller flocks. Essentially, they have got this capacity to live in high density or low density in a similar fashion to the way humans do and that is one of the factors that makes them extremely adaptable.'

When the bough breaks

In the dry summer of 1972–73, Budgerygahs [*sic*] were present in large numbers in the MacDonnell Ranges, and where water occurred. During December of 1972 flocks of 10,000 to 15,000 birds visited Ellery Creek Big Hole daily, wheeling around in tight formation and alighting to drink.

Hundreds settled on the water and drank with wings out-stretched. Those that stayed too long were unable to fly with their water-logged wings, and either drowned immediately, or fluttered to the shore or a rock surface, where they crawled out to dry. So great were the numbers alighting on trees that the weight broke branches of up to 40 cm diameter from Redgums.

—Dennis W. Chinner, *The South Australian Ornithologist*, 1977

The first record of talking parrots appears in a book written in the fifth century BC by Ctesias of Cnidus. This work, called *Indika*, was akin to an ancient Greek *Gulliver's Travels* version of India—without the satire. It included some fantastical tales of a race of one-legged people and unicorns that were brought by silk traders to Persia, where Ctesias was installed as court physician to King Artaxerxes II. Although the work was much scorned by later scholars, particularly for the race of people with feet so large they used them as parasols when at rest, Ctesias's description of the parrot rings largely true:

'The bird called bittacus has the tongue and voice of a man; its size is about that of a hawk, its head is red, and it has a black beard,' Ctesias wrote. (It is thought Ctesias misinterpreted the description and meant 'beak'.) 'Its neck is the colour of cinnabar; and it speaks the Indian language just like a man. When it has learned Greek, it speaks that language also just as well as its native tongue.'

By the time Pliny the Elder was kicking around in the first century of the Roman Empire, parrots were evidently common enough that a minor nobleman could afford to lose a few to maltreatment. In Book 10 of his epic *Naturalis Historia*—a 37-book encyclopaedia of the knowledge of the day—Pliny notes that

> there are some birds that can imitate the human voice; the parrot, for instance, which can even converse. India sends us this bird, which it calls by the name of 'sittaces'; the body is green all over, only it is marked with a ring of red around the neck. It will duly salute an emperor, and pronounce the words it has heard spoken; it is rendered especially frolicsome under the influence of wine. Its head is as hard as its beak; and this, when it is being taught to talk, is beaten with a rod of iron, for otherwise it is quite insensible to blows. When it lights on the ground it falls upon its beak, and by resting upon it makes itself all the lighter for its feet, which are naturally weak.

That birds share the human capacity for vocal learning has always been deeply interesting to science, as Dr Jacqueline Nguyen of the Australian Museum Research Institute (AMRI) explains. 'There are very few animals in the world that have the capability for vocal learning. There are humans obviously, but only three groups of birds; parrots including budgies, songbirds and hummingbirds have full vocal communication.'

Nguyen, an AMRI research associate specialising in ornithology and palaeontology, is part of the Bird 10,000 Genomes (B10K)

Project—a major international genetic sequencing initiative—that recently confirmed that the three types of 'talking birds' inherited the capability of vocal learning separately.

'So, they didn't all inherit that from the same common ancestor, but vocal learning evolved separately three times in the bird evolutionary tree. This is a good example of convergent evolution, where animals evolve a similar solution to a similar problem,' Nguyen elaborates. 'If you think about bird flight and insect flight, for example, these groups are not at all closely related, but they have both evolved wings to solve a similar problem. It will be really interesting to trace how vocal learning evolved right down to the species level. That is something that is in progress now.'

Recent studies have shown that budgies can grasp the basics of syntax or acoustic patterns in snippets of songs in much the same way humans learn grammar. When they are deprived of the chance to learn the full range of budgie-speak through their parents and kindred flock, captive birds will copy human speech and other sounds in their environment. 'They kind of perceive human speech as a kind of warble so they are able to match human speech like they would match songs in the wild,' Nguyen says.

But does this necessarily mean your budgie understands you if you say, 'I'm going out now' and it responds, 'I love you', 'Take your gas mask, mama' or even 'Bye'? Probably not!

Dr Robert Dooling, professor emeritus with the University of Maryland's psychology department and its Neuroscience & Cognitive Science Program, is one of the world's leading researchers of comparative psychoacoustics and one of the very first, if not the first, to use budgies to study auditory processes. He began studying them back in the early 1970s in St Louis, Missouri, at the city's Central Institute for the Deaf.

'They had a lot of conventional tests that they were then doing on chinchillas primarily, but also guinea pigs and the traditional sorts of auditory animals,' he says. 'I wanted to know . . . if mammals do it this way, do you suppose everybody does it this way? Or do birds do it differently? It was easy enough to buy budgies from a pet store and we worked out a way to frame them and test them. So that started for me a whole career of studying hearing and vocalisations, vocal production and vocal learning in budgies in a laboratory setting.'

Over the years Dooling and his colleagues have produced well over 100 scientific papers arising from the 'budgie business'—to say nothing of published abstracts. Papers such as 'Speech perception by budgerigars', 'The perceptual foundations of vocal learning in budgerigars' and 'Perception of complex sounds in budgerigars with temporary hearing loss' do not make for light bedtime reading, but this work has important implications for human hearing and vocal learning.

Dooling and his associate researchers were, for example, at the forefront of the discovery that birds can regenerate the sensory (hair) cells in their ears and can recover their hearing after it is damaged by noise exposure.

'That has all kinds of health implications,' Dooling says. 'If you could somehow turn on hair-cell regeneration in the inner ear in humans, you could cure 95 per cent of deafness in the country.'

Even after all these years the budgerigars, which in Dooling's lab are subject to behavioural studies rather than those that necessitate their death or maiming, are still a source of surprise.

'I remember one year, long after we had finished doing some

sparrow work, we were cleaning the birdcages and all of a sudden this male budgie produces this perfect swamp sparrow song,' Dooling chuckles. 'It was a perfect rendition and I can even tell you which bird it was from—this was from a sparrow that died two or three years prior. [The budgie] remembered it all this time and it was perfect.' To what end? No one is entirely sure, but Dooling guesses it was the budgie equivalent of a bloke puffing out his chest. 'It's always a sex thing, right? My guess is males are showing off, saying, "I have got all these talents. I can do this and I can do that."'

He has more than a little admiration for his feathered subjects. 'I have always enjoyed working with them. They seem like they have a relationship with you. When you get them tame they really seem to get you. Among birds—and crows are up there too—there is something quite personal and personable about a budgie.'

More recently Dooling has concentrated more on budgerigars' capacity to learn the warbles and calls that characterise the species.

'People think that language and thinking that goes along with language is what separates us from animals, but one of the fascinating things about birds is that they do some of the things that we do. They have to learn these vocalisations and they use them in a social context. It is the human and these 5000 or so species of birds that do this,' he enthuses. 'How does the smaller, simpler budgie brain do that? Then, you have songbirds with brains structured slightly differently from parrots, so what is that about?'

For a scientist, Dooling says, this is nothing less than a 'gold mine' for the study of genetics and language. 'It will be one of the stories of the future of how we unravel human language, what makes it work and what makes it fail, and it will come from this bird business, I think.'

Chapter 8

Ship ahoy!

IN 1841, WHEN Scottish journalist Charles Mackay published his *Extraordinary Popular Delusions and the Madness of Crowds*, he focused on the Dutch mania for tulips back in the seventeenth century. But had he been looking for more current and accurate examples of crowd zeal for his bestselling book, he might have popped down to the bird markets of Shoreditch or any major dock.

Budgies, then variously called shell parakeets, love birds, undulated grass parakeets, shell parrots, scallop parrots, splendid grass parakeets, warbling grass parakeets, zebra parakeets, zebra grass parakeets, beauregards and even canary parrots, were by far the most common passengers on vessels returning to England from Australia by the late 1840s.

It was not an easy passage for man, woman or bird, and more wildlife died than survived early journeys. Some creatures like koalas—despite almost 100 years of attempted exportation—did not actually make it alive to English soil until 1880.

But budgies are for the most part incredibly hardy, and this, coupled with their adaptability and general affability, ignited an extraordinary boom.

 ## *Birds, birds, birds*

Birds of all kinds have been very plentiful this season, and as a consequence an unusual number has been exported to England. Shell parrots especially have been exceedingly numerous, having been sold as low as 4d. [pence] to 6d. a pair. We heard of one gentleman who took home with him as a speculation as many as 10,000. Every vessel which leaves the Port takes with it large aviaries filled with birds of various descriptions. South Australia is absolutely establishing a new item in the export list.

—*The Adelaide Observer*, 22 February 1862

There was money to be made by everyone: birdcatcher, wildlife dealer, farmer, innkeeper and even, it seems, the lowliest steerage passenger.

In a published account of fire which broke out on the clipper *Orient* weeks after it left Port Adelaide in November 1861 an unnamed (but evidently male) passenger observed: 'The scene on the poop was striking, and if I may say so picturesque. The women from the second cabin and steerage were lying about in every direction; groups of three and four rolled up in blankets, and surrounded by bundles, baskets, watercans etc, etc. I even fancy I observed small cages of shell parrots ready to be taken to the boats.'

Writing in his *Handbook to the Birds of Australia*, published in 1863, Gould notes how his little Warbling Grass-Parakeet had taken off since the first importation in 1840. 'Since that period nearly every ship coming direct from the southern parts of Australia has added to the numbers of this bird in England; and I have more than once seen two thousand at a time in a small room at a dealer's in Wapping. The bird has also bred here as readily as the Canary.'

In August 1865, Anton Herman Jamrach—the stepson of the world's most famous wildlife dealer, no less—was to be found in Adelaide hunting up stock. A small advertisement in the *Adelaide Daily Telegraph* declared him 'open to buy Australian birds and animals such as kangaroos, etc' either on board the ship *Coonattoo* from 8 a.m. to noon or, propping up the bar, at the Port Admiral Hotel between 8 and 10 p.m. No doubt Jamrach Snr hoped the presence of a family member on board would help safeguard his investment. Even among the incredibly hardy budgies, the mortality rate was still as high as 20 per cent, packed, as they were, like sardines for up to three months.

 ## A whirling green cloud

[A] twittering of birds overhead breaks the silence of the almost deserted street. Until now I had forgotten that hereabouts dwells the great Jamrach. Sure enough it is from one of his shops that the clear bird melody comes, and, catching sight aloft of a flock of tiny green forms clinging to the wires of the window-frames, I cannot resist the temptation of a passing call. Young Mr. Jamrach, as usual, is very pleased to take me round, and I am equally pleased to follow him through that queer bit of Old London known as Jamrach's . . .

But I have one very strange treat in store at the top of the creaking old staircase. I say nothing of the rare and costly objects filling room after room in this huge curiosity shop, furnished piecemeal by many travellers from distant lands. But Mr. Jamrach, at the top of the house, unfastens a padlock, and we enter a room about sixteen feet square. He shuts the door quickly behind us. We are suddenly stormed upon by half a gale of wind, and there is a bewildering flutter of wings. We are in the midst of a whirling green cloud, for nearly two thousand pairs of pretty love-birds (the parakeets of South Australia, there called budgerigars) have risen *en masse*, startled from their resting-places.

After a moment the deafening twitter which had momentarily ceased is resumed, and some hundreds of the birds settle again like a swarm of bees on a shelf, and thick on all the ledges in the apartment. Tiny feathers descend upon us like snow. Mr. Jamrach has just taken over three thousand pairs to the Continent, and besides the two thousand pairs to whom the entire freedom this room is given there is another thousand lodged in cages about the premises. Healthy, hardy, happy fellows they must be, since they have room for flight, and eat something like four sacks of canary seed every week.

—'An explorer', describing a walk around London's East End,
London Daily News, 2 January 1885

The excellent Karl Russ offers the following description by way of a farmer called Elbert from Danzig, who lived for some time in Australia:

They are taken in hundreds, sometimes even thousands, aboard ship and put into comparatively small cages, which are arranged so that the perches are like steps of a stair, one above the other, in order to lodge in the smallest space as many birds as possible. These transport cases, closed in on three sides by planks, and on the fourth covered with wire, are usually completely filled. One row of heads looks over the other, and notwithstanding the close confinement, in spite of the frequent concussions, there is neither biting nor quarrelling, as is usually the case with other birds. In spite of their being crowded together like this . . . only an average of a tenth to a fifth of them perish.

But accidents could and frequently did happen, with 600 birds drowned when a single crate went overboard while loading at Adelaide's Semaphore jetty. Sometimes a shipment arrived with as few as 5 per cent of the birds lost, but other times infections and disease had wiped out the entire cargo before or shortly after reaching shore. Early on dealers learned the hard way that it was better to offload the little livestock quickly at a reduced margin, and prices fluctuated accordingly.

Trade escalated with the advent of faster steamers. The mail steamer *Souchays* returned to England in 1867 laden with copper, wool, flour and 15,000 pairs of budgies.

The fabulously named Lycurgus Underdown, publican of the Hindmarsh Hotel in Adelaide, was not a man to miss an opportunity. He amassed a huge stock of birds, which he advertised to ships' captains, and the aviary itself became an attraction mentioned in dispatches.

 ### Calling all captains

The stock of birds belonging to Mr. L. Underdown, at the Hindmarsh Hotel, is well worthy of a visit, and forms both a novel and interesting sight. The most prominent are two long cages containing about 35,000 pairs of beautiful shell parrots . . . There are also no less than 1,000 pairs of the well-known cockatoo parrots [cockatiels], besides small numbers of white cock-atoos, magpies, and the much-admired Port Lincoln Parrot [Australian ringneck]. A beautiful and delicately marked little bird, of which Mr. Underdown has about 7,000 pairs, is the zebra finch . . .

We are informed that already the proprietor has sold 9,000 birds of various kinds for exportation, most of them being purchased by captains of vessels going to England, to whom no doubt the establishment will prove a great convenience, while probably, with care and management, it may be made exceedingly profitable to the enterprising bird-dealer. The total number of birds in the yard would not be far short of 50,000 pairs.

—*The South Australian Register*, 28 December 1867

By 1868 the contents of Underdown's aviary were to be found on the deck of the clipper *Darra*, ironically because of a shortage of canary seed, which Australia then imported. As a South Australian pioneer who travelled aboard the *Darra* recalled:

There could have not been less than between 30,000 and 40,000 birds in large cages on the main deck. The season having been a very favourable one for securing large numbers of these birds, particularly shell parrots and diamond sparrows [diamond fire-tails], Mr Underwood had overstocked, and it was evident that the supply of canary seed would be absorbed. To meet the case, he made arrangement with Captain Lodwick to take the whole of his stock of birds to England.

In the first six months of 1879, more than 50,000 pairs of budgies were landed in England, with a single London dealer selling 14,800 pairs in four months. Birdmen with little ponies and carts piled high with budgies, canaries and Java sparrows became part of the London streetscape.

Back in Australia the demand for canary seed for feeding the captured birds was so large that it was imported in 200-gallon or 400-gallon iron tanks, with some export vessels requiring over 3 tons for feeding the birds in transit. This naturally gave rise to a secondary industry: a Mr Deslaundes became the first to sow a test acre of choice canary grass on land on Lefevre Peninsula, South Australia, for the little émigrés in 1880.

By 1885, with bird morbidity statistics down, the trade in budgie futures was up. Budgies had become mum-and-dad investments, with birdcages a common addition to steamer trunks on voyages back to the old country.

 *The budgie bubble*

By the SS *Taormina* a speculator is taking to England a large number of live Australian birds for the European market. They comprise about 5,000 pairs of shell parrots, a huge cage of zebra finches, with sulphur-crested and other large cockatoos. It is stated that Australian birds fetch a good price in London and Paris, and considering their small value here it is not difficult for a man to clear his passage to England and back, with a good margin of profit, by taking a few cages of small parrots.

Of course for the venture to be very successful he must have a certain amount of luck, but with the cages properly ventilated and the birds well looked after there would not generally be a death-rate of more than 20 per cent.

There is often a difficulty in getting steamers to take birds in large quantities owing to the deck room they take up, and even when they do the exporter has to arrange all sorts of contrivances in order that the cages should be fixed safe from damage by heavy weather. Notwithstanding all the difficulties, however, the bird trade is gradually increasing, which is after all the best test of its profitable nature.

—*The South Australian Weekly Chronicle*, 24 January 1885

In such a budgie-ish market something had to give, especially when breeding began in earnest in England and on the Continent. The retail price for wild-caught Australian birds dropped from £5 a pair in 1855 to as low as 3 shillings in times of glut. Writing 23 years after his visit to Australia, Anton Jamrach noted the decline in the market but saw the upside in the accessibility to lower classes.

 Australian love bird

It is not so many years ago that the fancy for the beautiful-plumaged birds with which our colonies are peopled could only be indulged by the very rich, but now, thanks to the great development of rapid steam communication with all parts of the globe, our houses can be made gay with the bright colours and graceful shapes of these sweetly pretty creatures at comparatively little expense. The scientific name is *Melopsittacus undulatus*, Undulated Grass Parrakeet. The birds are however, also known as Shell Parrots, Australian Love Birds, Zebra Parrakeets, and Beauregards.

They live in immense flocks all over Australia, and in ordinary seasons especially in the neighbourhood of the River Murray. Here they can be met with during breeding time by tens of thousands, enlivening the tall grass and the immense eucalypti with their continued fluttering and chirruping; the young birds are easily picked up from the ground and taken from the trees by the hand; and it is here where the Australian birdcatchers get all the Budgerigars that come to us from the Antipodes. They fill their bags of evening with birds, and on reaching home put them up boxes, holding one or two hundred each; these are taken to Adelaide and are sold to captains of vessels on the point of leaving for London, or they are sent to Melbourne and Sydney, and reach us via these ports.

Many a small fortune has been made in former years by the enterprising skippers, who brought large numbers over to England, which they sold here in the docks at £1 to £1 10s a pair, and which cost them out in Adelaide about 1s 6d each. One ship would often bring from five to fifteen hundred pairs, for which in the early days of the fancy there were always eager buyers. The first pair of Budgerigars ever brought alive to England was purchased for £22—about twenty-six years ago; since then the importations have yearly increased, and the prices proportionally been reduced until recently they could be purchased retail at 8s per pair.

The plumage of the male and female Budgerigar is exactly alike; but what may not be known so well is the means of distinguishing the one from the other. The only external difference of the male from the female bird is in the colour of the cere at the root of the upper mandible, which is a dark blue in the former, and brown or greyish in the latter. Imported birds begin to pair October and

November, which period corresponds with early summer in their native home.

Mr A. E. Jamrach

—*The Greenock Advertiser* (Scotland), 14 February 1881

Even though retail prices had dropped, demand for budgies remained very high and there was money to be made across the supply chain. Good catchers could make as much as £250 a year with their nets and snares.

South Australia was the hub of the bird trade and John Foglia of Rundle Street, Adelaide, was top cocky of the dealers. Between 1885 and 1909 he shipped, by his own estimate, an average of 40,000 birds a year. This did not include birds he supplied to the Australian market, and the animals in which he also dealt.

Many, many thousands of these birds were budgies. One year Foglia sold 16,000 shell parrots to one dealer, observing 'it took three of us three days from 7 o'clock in the morning to 7 o'clock at night to count them'.

In 1905 he paid £65 to secure 11 tons of room on the top deck of the German boat *G.M.S. Rhein* for 14,000 birds most of which were budgies (fetching 2 shillings and 6 pence a pair on the European market) and zebra finches (between 1 shilling and 3 pence and 1 shilling and 6 pence), but also included diamond firetails, Major Mitchell's cockatoos, corellas, pigeons and wholly protected Cape Barren geese, which were exempted from export prohibition because Foglia provided sworn testimony they were born in captivity.

Foglia told one of the newspaper correspondents, who often visited his shop, that he expected to sell most of the cargo—including a number of kangaroos, wallabies, bandicoots and possums—in

Antwerp. Two years earlier Foglia had taken 17,000 birds to Antwerp and proudly asserted that 'only 200 died on the voyage and I maintain that number would have died if they had been wild in the bush. We reckon generally on losing the most birds between here and Fremantle. It is the worry they undergo in being shipped that kills them. They settle down after a few days at sea'.

Not so lucky were the Australian ringnecks—then known as Port Lincoln parrots—because of the 'injurious effect sea air has on their eyes, which often turn white and fall out on the voyage'.

When nature lays the table

The budgerigar is an occasional visitor to South Australia. It is called Shell-Parrot or Canary-Parrot by the colonists of this country with whom it is very popular. One of the favoured breeding-places, the object of my close observation, is Mallee Shrub, a beautiful eucalyptus wood, extending to the Murray, from its mouth to the first big curve of the river. If, in this inhospitable region, after a wet winter, it continues to rain abundantly in the spring—that is, at the end of September—grass will grow to an unusual density and height. Many square miles, otherwise unmistakably bearing the stamp of a dreary sandy desert, are suddenly covered with the finest kangaroo-grass which, under the influence of the summer heat, shoots up to the height of three feet. The blossom develops rapidly, and about five or six weeks later runs to seed. Long before, however, the pretty parrots have appeared in innumerable swarms, and now they apply themselves eagerly to their task of nesting operations.

The remarkable shape of the Mallee is particularly favourable for the purpose of nesting. About eight stems grow out of the same roots to a height of about 36ft with white bark and scanty tops.

Every hollow trunk, every knot-hole, in case of necessity even every suitable cavern in the roots is used for nesting, often by two and three couples together. In a few weeks things get quite lively. The ripe seeds of grass are perfectly suited to feeding the young. Anyone who happened to pass such a spot at this time would be able to catch hundreds of them easily in his own hands. In huge swarms they fly up from the grass when they hear anyone approaching. They perch in long rows on the bare twigs, amusing themselves by chirping songs and unsuspectingly they watch blood-thirsty man raise his gun to let fly a charge sufficient to bring down dozens at once. At last the available seeds are consumed. Perhaps there is also a lack of water. The passion for travelling seizes the birds and leads them further. Their next aim is to reach Alexandrina and Wellington lakes, through both of which flows the Murray before it discharges itself into the sea. It is uncertain whether the morasses there provide them more abundantly with grass seeds or whether the proximity of fresh water attracts them; but this is the spot to which every year the bird-catchers come in order to lay their snares and to catch many thousands. This, however, only applies to the years with abundant rainfall. But, in other years, when the rainfall remains below the yearly average, budgerigars seem to disappear completely. No doubt they move to the far north where often, in the summer heat, violent thunder showers occur, where, as stated before, they completely change a sandy desert in a short time into a grassy plain. All migrating parrots seem to know that by anticipation; for where nature has laid their table, they attend.

—Adolf Engelhardt, in Karl Russ (trans. M. Burgers), *The Budgerigar: Its natural history, breeding and management*, 7th ed., 1928

Chapter 9

Feathers fly

THE BUDGIES THAT survived the sea voyage and found themselves being cooed at in cages in colder climes were arguably the lucky ones. The little parrots were often ill-used both in their native land and abroad, being variously dished up as pie, used as quarry for shooting matches and even used to dress hats.

Parrot Pie was a popular colonial dish which featured in later editions of *Mrs Beeton's Book of Household Management* in the 'Australian cookery' section. It called for '1 dozen paraqueets, a few slices of beef [to line the base of the pie dish], 4 rashers of bacon, 3 hard-boiled eggs, minced parsley, lemon peel, seasoning and stock', and a topping of puff pastry.

Others made sport of the birds, with South Australia's Hamley Gun Club playing host to the governor of Tasmania, Sir Charles Du Cane, for a 'capital afternoon' of shooting budgerigars in December 1870. The budgies apparently saved the day as the pigeons were just not up to mark, while 'the shell parrots afforded fine amusement both to the trap shooters and those who were able to wipe their eyes'.

'The shells make excellent shooting,' an unidentified correspondent wrote of the club event in *The South Australian Chronicle and Weekly Mail*:

> *They go away very swiftly, and show great variety in their modes of flight—some dart straight from the trap like a flash of lightning; some mount like a rocket and suddenly pouncing downwards, skim along the grass; some fly straight like quail; others zigzag after the fashion of snipe, or wing their course upwards; some go directly from the shooter, many make for the muzzle of the gun; and there is a fair proportion of right and left-hand shots.*
>
> *This season they can be got in great numbers, at very cheap rates. The best-sized shot for them is dust, though many persons prefer No. 10.*

Incidentally, His Excellency the governor came second by a feather to his host J.I. Stow, QC, when the combined shoot-out of both budgies and pigeons was tallied.

As late as 1882 the South Australian Gun Club was recorded as substituting budgies for pigeons at its shooting ground, the Morphettville Racecourse. The protection of native fauna was still a way off, but what would give the cause great impetus was fashion.

New York's Metropolitan Museum of Art holds in its collection a silk, lace and velvet bonnet topped with three whole dead budgies. Dating from 1890, it was gifted to the museum by philanthropist Susan Dwight Bliss in the 1920s. It is a fine example of 'murderous millinery', which became the wings beneath one of the earliest conservation movements.

In many ways the demand for budgies was a by-product of a general frenzy for feathers. The acquisition of rare, exotic and beautiful creatures from far-flung places was part and parcel of nineteenth-century colonial imperialism which, overlaid with quirky Victorian era sensibilities and more than a little flight envy, saw birds simultaneously admired, hunted, plucked and taxidermied en masse.

London was the epicentre of the world feather trade which, at its peak, saw more than 1.5 million pounds of plumage (680 tons) go through its Mincing Lane auction houses a year. A single order placed by one London dealer in 1892 for 6000 birds of paradise and 40,000 hummingbirds underlines the scale of the slaughter during the 'plume boom'. The most exotic, rarest and colourful birds were most at risk, and many millions of birds were killed around the globe—including in Australia, where plume hunters threatened the existence of the emu, lyrebird, egrets and bowerbirds.

The emu, once seen fit only for mats, hearth rugs and quirky occasional table legs by colonists, began to feature in the London social pages as an adornment for the richest and most fashionable. At a 'drawing room' (a formal party) held by Queen Victoria in May 1891, a Miss Mabel Riddell presented as a vision in white chiffon, with basque bodice and chiffon skirt both fringed with emu feathers: 'Ostrich feather tips formed an aigrette on one shoulder and emu feathers on the other, the latter looking very light and charming.'

The more exotic and flamboyant headdresses, capes, drapes, fans and flounces reflected the wealth and status of the wearer, but even the lowliest housemaid could aspire to a bright wee twist of budgie feather to tuck into her hat on a day off.

Measures introduced by New South Wales, Victorian and South Australian parliaments to protect birds in the early 1880s and 1890s had done little to curtail the killing, and some put the welfare of imported game like pheasant, partridge, grouse, Californian quail and white swans ahead of native birds' safety. 'An Act to protect certain imported and other birds' was how legislation introduced by the New South Wales Parliament in 1881 was framed.

But as more and more species of bird were pushed to the brink, people began to take notice. In 1894 the first Australian branch of the Society for the Protection of Birds was formed in Adelaide and included a good number of influential matrons who, in accordance with the rules of the organisation, eschewed the wearing of feathers.

By 1897, when the group made an impassioned plea for bird welfare in a deputation to the state government, there were 407 members. At the time the Society's patron was Lady Victoria Buxton, the wife of the South Australian governor who, along with her daughters, had publicly abandoned egret feathers. The deputation included in its case the budgerigars 'taken to England by the masters of ships ... the birds were often crushed in a small space, where large numbers of them died on the voyage'.

Their efforts were rewarded with the introduction of the *Birds Protection Act 1900* (SA), which for the first time offered the budgie some safety for at least part of the year. The introduction of a closed season in South Australia, which prohibited the trapping and exporting of budgies between 1 July and 12 January, had a significant impact on trade in that state.

By 1914 our forthcoming friend Foglia, who once used to make between £500 and £600 out of a trip to Europe, had been forced to switch his attention to American markets. 'The other states have no such stringent laws and in consequence they have all their birds caught, boxed and shipped before we in South Australia have ours caught,' Foglia explained. 'They can catch practically all the year and in consequence have the Continental market full before any birds from this state can arrive.'

This had a flow-on effect to the professional catchers operating in South Australia, who were now lucky to earn £70 a season. Accordingly their number had dropped from ten to two.

But, while Foglia apparently observed the rules, there was no policing of borders and no shortage of less scrupulous smaller dealers. With gaping holes in the legislation between the states, vast flocks of native birds could, and did, pass through.

Newspaper listings show that among the birds of paradise, osprey, vultures and peacocks, the London Commercial Sales Rooms was still offering twenty boxes of lyrebird tails and over 7000 parrots at a time in 1907.

 ## Extermination of plumage

With regard to the cruelty practised on the egret by the plume-hunters, and as showing the havoc wrought by these barbarians reference may be made to a paper read recently before the Royal Colonial Institute by Mr Jas. Buckland, who observed that the lyrebird of Australia is now almost extinct. A few years ago 400 were killed in one district alone in a single season to supply the plumage market. Last year only 52 tails were catalogued in the London feather sales. The egrets of Australia are being exterminated at such a rate that the best plumes now fetch £8 per ounce in London. Mr Buckland mentioned that the egret feathers are obtained by slaughtering the parent birds at the time when they are hatching their young. The emu has already been exterminated in Tasmania, Victoria and South Australia, but it still lingers on in some places and last year 1019 emu skins were sold in London at 17s 6d each. The bower bird, which is famous for the artistic manner in which it decorates its nest, has been so killed out that last year only 21 could be had in all London. They realised 1s 7d per skin.

—*The Belfast News-Letter*, 15 April 1911

In the long term it would take the grassroots environmental advocacy of organisations like The Gould League of Bird Lovers, founded in Australia in 1909, to help bring about changes in attitudes to native fauna.

Farmers regarded not a few native bird varieties as pests and were among those who were unhappy when the New South Wales chief secretary revoked the five-month February-to-June open season on the 'Warbling Grass Parrot (Budgerigar or Love Bird)' in March 1922—but for altogether different reasons from the bird-catchers, traders, dealers and exporters. Announcing the bird would henceforth be 'absolutely protected', the government spelled out for farmers' benefit that 'these birds do not cause any destruction to wheat but live mainly on grass seed'.

The action was necessary because of the great numbers being captured. 'Reports show that in the Byrock district [in Bourke Shire] alone some twelve to fourteen thousand have been captured recently and sent abroad. The depletion of the flocks of these birds in such a wholesale manner would, it was claimed, lead to almost total extinction.'

But by then budgies had already begun to put their own brake on the lucrative export business—and it most certainly wasn't because they were dying out.

Chapter 10

Mon petit parakeet

EVEN AS THE Earl of Derby had fussed around a nesting hole in 1848, anticipating the emergence of the first budgerigars bred on British soil, bird fanciers across the Channel had been breathing down his neck.

It was obvious that, if the birds could be bred where the market was, it would save a good deal of time, effort and money. Avian fatalities would be greatly reduced—and, possibly, even deaths of stressed wholesale buyers. Our expert contemporaneous chronicler Russ tells the sorry tale of a Mr Bolzani, who attempted the first wholesale import of budgerigars into Berlin in 1850. Bolzani was in London buying seashells, they also being quite the thing in the nineteenth century, when he was offered 500 pairs of birds just landed from Australia 'cheaply'. Unfortunately, by the time he got them home all the birds had died 'and he was unable to sell a single one'.

By the end of the 1860s several large commercial breeding farms had been established on the Continent. The first of these was in Belgium where the Flemish were already well practised in canary breeding, having produced several cornerstone types from the Old Dutch variety over the course of a century.

Canaries might be useful to save miners' lives, but naturalist and founding member of the German Ornithological Society Dr Karl Bolle apparently believed the budgie could bring light to the whole working class.

 Marx my words

From the moment when the canary-breeders combine with the upbringing of those golden-coloured songsters the breeding of other as yet expensive domestic birds, their profit will be doubled. Many a miserable hour of drudgery at the loom and in the glass foundry would be brightened, many a tear of poverty would be dried up in that way, and in a short time perhaps a great number of these charming creatures until now only found in the houses of an opulent minority would be the common property of the people. For the purpose of such breeding no bird could be more warmly recommended than the Budgerigar.

Everyone who lives in or near a town of any size knows that extremely lovely parrot from Australia, which has among many other points in its favour the advantage of breeding even in captivity with extraordinary ease and of bringing up young in unusual seasons, and in spite of its much warmer native country. This explains the increase in general popularity of Budgerigars as domestic birds in the course of the last decade.

— Dr Karl Bolle, 1859, in Karl Russ (trans. M. Burgers), *The Budgerigar: Its natural history, breeding and management*, 7th ed., 1928

As budgie breeding businesses were established in Amsterdam, Paris, Cologne and the zoological gardens of many major cities, keepers discovered just how prolific budgies could be. This, as Bolle

77

had anticipated, pushed the price down on the Continent as the Australian bird began its ascendancy over the canary.

The budgies seemed to find life particularly agreeable in the south of France where they established a stronghold in Toulouse, 'The Pink City', in the 1880s. The Toulouse goose, a very old breed of placid domestic goose known as the producer of foie gras, was fatly famous as top bird of the region when the little Aussie chick began turning heads.

A fairground bird dealer named Bastide was among the first to start breeding the *perruche ondulée* (budgerigar) in Toulouse. A visitor to Bastide's home in Avenue Frizac described how the dealer had transformed a quadrangle into a huge tiered aviary comprising 48 shipping-container–sized cages housing more than 15,000 budgerigars, with breeding pairs at the ground level and young birds in the upper ones.

It was the autumn of 1888 and 'everyone in the establishment was busy separating the cocks and the hens for winter; they were cleaning out the nest-boxes and the quantity of clear [infertile] and unhatched eggs were so enormous that they were taken away in the wheelbarrow full,' J. Bailly-Maitre recalled in *L'Oiseau*, the French ornithological magazine, in 1925.

Two years earlier another breeder had set up just down the road and when the writer returned to the district in 1913, he found Blanchard's Ornithological Establishments was also producing tens of thousands of budgies annually. Meanwhile Bastide's sons had taken over the original business and it had become a venture employing 20 people and including a huge 1000-square-metre open-air 'Jardin d'Acclimatation'.

'The average population of Bastide's Perrucheries is between

80,000 and 100,000 and if one were to place the 290 aviaries end to end in a straight line there would be 1200 metres of them. Six hundred, eight hundred or even one thousand sacks of millet are required by the Bastides to feed these birds,' Bailly-Maitre marvelled.

Toulouse was clearly Budgie Central, but then came the Great War. It devastated the local breeding industry, as Bailly-Maitre reported:

> *Though most industries suffered on account of the war it can safely be asserted that none lost more heavily than the budgerigar breeding establishments. England, Germany and Russia were their chief customers. Suddenly all those outlets closed and besides it was necessary to suppress all useless mouths. Consequently, after August 1914, the House of Bastide had to slaughter all the 120,000 budgerigars in their aviaries at the declaration of war.*

Other bird breeders and collectors endeavoured to keep their stock alive after war broke out. The famous French–American ornithologist Jean Delacour had built up a magnificent collection of birds, including hundreds of budgerigars, at his family estate at Villers-Bretonneux, the scene of some of the fiercest battles of World War I. From September 1914 to August 1918 four major battles, including the Battle of the Somme, were fought in the region.

Delacour later recalled in *L'Oiseau*, the magazine which he founded in 1920 and edited until World War II, how the battle lines ran through his aviaries while he himself served in the French Army:

> *In September 1914 the Germans reached Villers-Bretonneux, killing and letting out a certain number of my birds, but the bulk*

of the collection was saved and, after the retreat of the enemy, kept in excellent condition for nearly four years. In March 1918, unfortunately their last offensive brought the Germans back to our grounds, where they were stopped first by British and French troops, and later on the Australians. Severe fighting took place for several months in the park, the result being the complete destruction of the birds and everything else there was. All my birds were killed, all the trees and plants destroyed and all the buildings were wrecked ... Where the gardens, conservatories and aviaries had been, only tangles of broken steel, glass and wood remained.

 ## Parrots detect Zeppelins

In England recently a pair of budgerigars, the beautiful Australian parrots, imitated the geese that saved the Capitol of Rome. The parrots warned a household that a raiding Zeppelin was overhead.

This story is told by Mr. W.P. Pycraft in 'The Illustrated London News'. He writes:—

'I had the misfortune to be out of London during the last Zeppelin raid. As I was away at the seaside, I induced a friend to take charge of a pair of budgerigars belonging to my little daughter. The cage was hung in his garden, just outside his bedroom window. Hearing a sound of most exciting chirping and fluttering, he got out of bed to ascertain the cause, and at once discovered a Zeppelin almost overhead. Aeroplanes, at any rate, where they are not too common, produce like alarm.'

—*The Herald* (Melbourne), 10 December 1915

Chapter 11

Of peas and Japanese

THE APPEARANCE OF a canary-yellow budgie in an aviary in Belgium in the early 1870s was a light bulb moment in the development of the modern bird. Despite the hundreds of thousands of birds being imported into and bred on the Continent, this was the first recorded colour mutation of the species in captivity and it eventually helped change the breeding of budgies from happy accident to algebraic art.

A handful of years earlier a myopic monk who had twice failed to obtain his teaching diploma delivered a lecture at the Natural History Society of Brünn (now Brno) in the Czech Republic. His topic was the cross-breeding of peas in the St Thomas's Abbey garden.

After eight years of experimentation, growing and recording the development of 10,000 hybrid pea plants, Gregor Mendel had developed his Laws of Heredity and laid the foundation of modern genetics.

Mendel sent copies of his paper *Versuche über Pflanzenhybride* (*Experiments on Plant Hybrids*) to leading members of the scientific community and learned societies around Europe, where it was dutifully filed but not fathomed. Darwin was still everyone's darling and it took another 75-odd years before his Theory of Evolution was fully reconciled with Mendel's ideas as a coherent calculation in *Evolution: The Modern Synthesis* by Julian Huxley.

In 1900 Mendel's work was 'rediscovered' after three botanists, each working independently, duplicated the results of his pea experiments with evening primrose, cereal crops and hawkweed. But the budgie breeders were already well ahead of them—in practice if not theory—as they carefully selected birds in attempts to develop the yellow variety. In 1880 an aristocratic collector presented two yellow budgerigars at an exhibition in Berlin, establishing that colour breeding was also underway in Germany.

Nonetheless, when an Australian bird exporter named Gard captured two wild yellow birds in the bush, they were still sufficiently rare to be exhibited in the South Australian Court of the Colonial and Indian Exhibition—a huge 1886 showcase of 'every portion of Her Majesty's Empire' held in Kensington, London, that attracted over 5.5 million visitors.

 Rara avis

Mr Gard, a well-known bird exporter, sailed for the old country with a large consignment of shell parrots and zebra sparrows [zebra finches]. Upwards of 5000 of these feathered voyagers are contained in six large cages. The birds were as thick as bees in a hive, the perches being places about 3 inches apart, and each case held about 500 pairs. The exporter has sent home as many as 35,000 in one season, with an average death rate of less than 1 per cent.

Amongst them is a pair for which he asks £50, as the male parrot is a rara avis, being a shell parrot whose plumage is a beautiful yellow, as pure as the tint of a Belgian canary's. His mate is very yellow, but slightly streaked with green, and their progeny resemble the mother. Mr Gard was hoping to breed more of the pure yellow birds from them, but he has failed to do so.

The canary parrot is one of only three that he has ever seen amongst millions of the tribe. Mr Gard has been months collecting these birds visiting distant places northwards and obtaining a large number near Wilcannia. He states that he has captured most of them with a net and 'knows their habits better than they do themselves'.

—*The Leader* (London), 15 May 1886

By the time Gard landed his fluttering cargo, yellow birds were being bred in England from stock imported from Belgium. And, in 1900, the first selectively bred yellow budgies were sent back to Australia to meet the demands of local bird fanciers.

But if the market for yellow birds was strong, it was nothing compared to the frenzy that the appearance of blue birds provoked. Again, it was the Belgians who claimed the prize for producing a bird that briefly became the most coveted in the world.

The first blue budgerigar appeared in the aviary of a breeder named Limbosch in Uccle in 1878. This is now a leafy residential area housing some of Brussels' wealthiest folk but back then was predominantly rural. It's an area that, at the beginning of the twentieth century, would become more commonly associated with the development of a bearded bantam chicken known as the Barbu d'Uccle than the breeding of blue budgies. But Uccle is significant in terms of tracking the mutation because in the same year a second blue bird, this time a female, hatched in the aviary of another local breeder.

Karl Russ, who as founder and publisher of the magazine *Gefie-derte Welt* (*Feathered World*) was across the choicest ornithological news of the day, received the first report of this phenomenon from Louis Vander Snickt, a former manager of the zoological gardens at Ghent and Dusseldorf. In 1881, Vander Snickt wrote: 'No bird all over the world has caused so much admiration and the same time so many expectations as the blue Budgerigar of M. Limbosch.'

The Limbosch bird had a blue body and a white head, while the second Uccle bird was blue all over its body with a yellow head and tail. But the two birds were never paired, and it remained all quiet on the blue budgie front until another Belgian bobbed up with a pair, reported to have been bred from Dutch stock, at the Royal Horticultural Society's London exhibition at Lindley Hall in 1910. The following year a Mme Quentin de la Gueriniere exhibited a blue in Paris.

But many of the early 'skyblues'—as they became known in the ever-growing spectrum of budgie colours—were not strong birds and failed to thrive. The first blue budgies brought out to Australia for the Taronga Park Zoo by Ellis S. Joseph turned out to be duds, although world-renowned zoological collector Mr Joseph, who was yet to claim fame for the first live transport of a platypus to New York in 1922, was apparently quite a specimen himself.

 The platinum platypus and Ellis S. Joseph

Ellis S. Joseph is a mystery in himself. Imagine a big man, rather like Falstaff inflated at the local garage; height about 6½ft., weight 21st., as lithe and active as a midnight cat, as delicate with his fingers as a florist's assistant, with a voice like a big funnel rumbling steam, a hand-grip like a rat-trap, and the kindliest eyes that ever twinkled at a height of six feet six inches above Australia . . .

That Columbus of its tribe might have been called a platinum platypus. 'How did I get it across alive?' repeated Mr Joseph. 'It took me nine years to learn how to do it. It was my greatest ambition, and it cost me about £1000. A Sydney man first invented the tank in which the platypus was housed, and I added some improvements of my own. There was just one other reason for my success besides the tank. I turned into a platypus myself.

'Yes; on that trip I slept with the platypus, ate with it, drank with it, and lived with it. I had only one thought, and that was platypus.'

—*The Sun* (Sydney), 6 May 1923

Note: The platypus exhibited with fanfare in the New York Zoological Park was the only survivor of the six Joseph was given government approval to export to the United States. It lived for seven weeks, on display for an hour a day in New York, before it too died.

Of course, World War I had hugely interrupted the wildlife trade and animal breeding, as it had everything else. In Toulouse both the houses of Bastide and Blanchard would have to rebuild their respective businesses, the former concentrating on green and yellow birds and the latter turning its attention to other hues.

Back wandering through the 'Perrucheries Blanchard', again with Bailly-Maitre in the 1920s, it is impossible not to marvel at the resilience of both bird and man. By this time Blanchard had gone beyond the blue (of which he had bred hundreds) to produce lilac, olive, grey and jade, and was already hot on the tail of a white or albino budgie, as Bailly-Maitre reported in *L'Oisear* magazine:

> *When one visits collections of Budgerigars as important as those of Blanchard, one is struck by the numerous variations in the plumage of these birds. There are, we believe, few kinds of animals, which having been for such a short time in the hands of man, have produced so many varieties. One could name plenty of other species which, although reared in captivity for many generations, have not yet given us any one of the individual modifications of which the Budgerigar is so lavish.*

The world had not seen anything yet. Enter Hans Julius Duncker, a pioneer of the then fledgling field of avian genetics. The German ornithologist had already established that colour in canaries was a polygenic trait when, in 1925, a wealthy businessman and keen aviculturist bankrolled him to work with budgies.

Carl Cremer's deep pockets meant having access to birds with colour mutations as soon as they appeared and, through hundreds of test matings, Duncker was able to establish the Mendelian inheritance pattern to predict the outcome of crossing the twelve recognised budgerigar varieties of the day.

His work earned him a gold medal from the British Budgerigar Society and his original tables giving the (theoretical) outcome of 1830 different matings was the crib for breeders the world over, many of whom were helped greatly by Duncker's work when the Japanese dramatically entered the market.

Bird-keeping has a very long tradition in Japan. Song contests were held during the Edo period (1603–1868), and birds were trained to do tricks, setting the stage for Japan's enthusiastic adoption of the budgie in the twentieth century. The town of Sakai, on the edge of Osaka Bay, was particularly renowned for its devotion to cage birds, with more than 100 bird dealers and an estimated two-thirds of all households boasting Java sparrows, finches and budgerigars.

Henry Vollam Morton, a legendary British newspaper columnist and writer, recounted the Crystal Palace Bird Show of 1927 when a single bird in the skyblue, cobalt and mauve classes fetched between £50 to £175 each, while a pair could net between £100 and £500. 'We used to call that section of the show millionaire's corner,' one breeder told Morton. 'I remember how Japanese would come round to the breeders for blue budgies with wads of bank-notes—sometimes £2000—strapped round their arms in a leather case.'

Huge numbers of birds were exported to Japan, with dealers in Britain advertising to buy as many as they could for resale at mind-boggling prices—just imagine a humble budgie costing the price of a new car! It seemed the Japanese couldn't get enough of them . . . until they started breeding and exporting the birds themselves in huge numbers. Effectively flooding the market, the influx of birds lowered the price of blues and cobalts back down to around 10 shillings a pair by 1930.

The Americans were by now also well in on the act. In 1928, one breeder in Los Angeles was breeding and selling 15,000 budgerigars, including blues, mauves, whites and other colour varieties.

As the 1929 Wall Street crash reverberated around the world and the Great Depression hit, many European breeders were stuck with very many birds in hand suddenly worth a fraction of what they'd anticipated. And, to make matters worse, parrots—large and small—were about to hit the headlines for all the wrong reasons.

Chapter 12

Sick as a parrot

TO ANYONE WHO grew up in the 1970s, the sad tale of the American household that woke on Christmas morning to find a dead parrot under the tree might conjure a certain Monty Python skit. But on 25 December 1929 it was not funny . . . not funny at all.

This prostate Polly marked the start of an outbreak of parrot fever of an entirely different nature to that which had gripped the Japanese. This was a fever that had at least a one in six chance of killing people who contracted it from caged birds.

It put the panic in pandemic and generated improbable head-lines like 'Coroner's Warning Against Fondling Parrots, Grave Risk in Kissing Birds'. People literally ran past pet shops holding hand-kerchiefs over their noses and mouths. As Harvard history professor Jill Lepore put it in *The New Yorker* 80 years later, 'Before it was over, an admiral in the U.S. Navy ordered sailors at sea to cast their pet parrots into the Ocean. One city health commissioner urged everyone who owned a parrot to wring its neck. People abandoned their pet parrots on the street.'

A zoonotic disease is one that can be transmitted from animals to people. Psittacosis is a zoonotic disease that can be passed from parrots to people. Budgerigars are parrots. By the time of the outbreak, they were well on their way to becoming the most popular pet on the planet.

The link between caged birds and human contagion was made 50 years before that fateful day in 1929 that Mrs Lillian Martin of Annapolis, Maryland excitedly lifted the cover off a gift from her husband to find an ex-parrot. Five days later she fell gravely ill herself.

Swiss physician Jakob Ritter had more than the usual medical motivation to want to get to the bottom of a household epidemic that broke out in Uster, in the canton of Zürich, in 1879: the house affected belonged to his family. Within a few short weeks, three of seven people who had sickened, including his beloved bird-loving eldest brother Heinrich, were dead.

After noting symptoms that were characteristic of both pneumonia and typhoid, Ritter carried out the post-mortems himself. Finding his brother, the daily maid and the birdcage repairer shared abnormalities of the lung, heart and spleen, he then looked for a common cause. Ritter identified Heinrich's handsome wood-panelled study, with its floor-to-ceiling aviary containing twelve recently imported parakeets and finches, as the epicentre of the outbreak. The birds, he believed, were vectors for a condition he dubbed *pneumotyphus*.

Similar cases were recorded in Berne and Leipzig before an outbreak in Paris in 1892 saw 50 people stricken, resulting in sixteen deaths.

Subsequent investigations revealed all but three cases could be directly linked—either by handling or purchase—to a consignment of parrots from South America. In 1895 Antonin Morange, who was writing his doctoral thesis on the Paris outbreak and has apparently been remembered for nothing else outside his native land, called the disease for what it was, *psittacosis*—after *psittakos*, the Greek word for 'parrot'.

In 1929, just a decade since the end of the final wave of the 'Spanish flu' pandemic, health authorities were hugely sensitive to any suggestion of sudden eruption of disease. The deadliest flu in history had infected an estimated third of the world's population, killing an estimated 50 million people (three times the number who had died during World War I) including 675,000 Americans—enough, in fact, to temporarily slash twelve years off their average life expectancy.

The world at large had no defence to speak of against the unseen microbial enemy. Penicillin, the first naturally occurring antibiotic, had been discovered in 1928 but was still more than a decade away from being used to treat people, so the psittacosis outbreak was a very real cause of concern.

The 1929–30 pandemic began in Argentina where, over several months, 100 cases of unusual influenza were linked to a consignment of sick amazon parrots from Brazil. At least 59 of the birds had been hastily exported before authorities joined the dots.

Ultimately there were more than 750 cases reported worldwide, and an average of one in six patients died. The toll was highest in Germany, with 215 cases and 45 deaths. Lillian Martin, one of 185 cases in the US, made a full recovery, but 35 others died, including two bacteriologists, William Stokes and Daniel Hatfield, who were exposed to the intracellular bacteria—now identified as *Chlamydia psittaci*—during their investigations.

The Americans had acted promptly to close their borders to the menace. President Herbert Hoover issued an executive order on 24 January 1930: 'no parrots may be introduced into the United States or any of its possessions or dependencies from any foreign port.'

Over the following months Australia, New Zealand, England and many European countries followed suit. With the movement of parrots so restricted the number of cases of psittacosis dwindled, although some people remained susceptible to the mere suggestion.

Parrot fever: another case suspected at Leamington

Illness follows parrot's death

Taken ill shortly after the death of her parrot, Mrs. Brison, of Duke Street, Leamington, was conveyed to the Warneford Hospital, but up to the present doctors have been unable to diagnose her illness, although parrot disease (psittacosis) is suspected.

It appears that shortly before Christmas Mrs Frank Brison bought a parrot. She was dissatisfied with the bird because it would not talk, and shortly after Christmas she exchanged it for another, taking precautions to obtain a bird which was not newly imported. Within ten days of its purchase the second bird was found dead in its cage. Mrs Brison was very much upset at this, and a few days later she was taken ill.

She complained of trouble in her chest and her nerves were upset, and on advice of her doctor she was taken to Warneford Hospital, where she remains at the present time.

In an interview with her relatives this week a *Warwick Advertiser* representative was informed that Mrs. Brison had taken great care not to come in contact with the parrot. She did not touch it, and it was sprayed regularly with disinfectant. Her husband fed the bird at night.

> Mrs Brison's relatives were of the opinion that the illness is due
> to the shock of finding the parrot dead, following closely on from a
> severe cold from which she was just recovering.
>
> —*The Warwick & Warwickshire Advertiser and Leamington Gazette,*
> 15 February 1930

All fingers remained firmly pointed at the parrots of South America until December 1931 when three matrons from Grass Valley, in northern California, presented with 'toxic pneumonia' and died one after the other over a period of ten days.

The women, aged in their sixties and seventies, had last been together at one of their homes for a regular catch-up over coffee. After burying his wife and farewelling her friends, the home-owning husband had also become critically ill. The case was sufficiently worrying for public health authorities to refer it to the epidemiological experts at the Hooper Foundation.

The Hooper Foundation was the first medical research foundation in the United States that was incorporated into a university. Established thanks to a million-dollar endowment, it had a clear charter to investigate 'the nature and cause of diseases and in the methods of its prevention and treatment, and . . . disseminate gratuitously all knowledge so acquired'.

It was in the Foundation's laboratories at the University of California that some of the greatest advancements in public health in the twentieth century were made. Under the directorship of the brilliant and energetic Swiss-born Karl Friedrich Meyer, botulism, brucellosis, encephalitis, plague and psittacosis were all hard nuts cracked by the disease detectives at Hooper.

Alerted to events upstate, Meyer told the local health officers to look for sick or dead parrots. They duly reported back there was a shell parrot—or 'lovebird', as the Americans then knew the budgie—chirping away in a cage hanging near the kitchen banquette where the women had last happily sipped and chatted. A second bird had died on 1 December and been buried in the garden of the bungalow. Meyer ordered its disinterment and for the little corpse to be brought to the San Francisco labs along with the surviving bird.

Within hours Meyer had established the birds were carrying psittacosis, and the parasite had also been isolated in pathology samples from the fourth victim. Meyer hit the road to drive the two and half hours to Nevada County Hospital to interview the dying man about the origin of the birds. With his last conscious breaths he explained the birds had been bought from a travelling peddler. At Meyer's urging police tracked down the man, and under threat of penalty unspecified he revealed that the birds had been reared in a backyard in Los Angeles. The inference was clear: psittacosis was present in the locally bred parrot population.

Grass Valley, as home to some of the oldest and richest gold mines in California, was faring better than many other places in the United States at that time. The mines remained hugely productive throughout the Depression, employing 4000 men. It was the kind of place where a man down on his luck could knock on the door and hope to trade a couple of birds for a dollar or two. Health authorities would soon discover the peddler was not alone in trying to make a living this way as more cases of psittacosis were reported.

Sunny California is not as hot as central Australia, but it was definitely more conducive to budgerigars than much of Europe. The birds were so hardy and such good breeders that you hardly needed to know anything about them to produce a saleable product. This is what hundreds of people were doing to try to make a buck. Authorities discovered budgies had become a whole backyard industry, with a lot of birds being produced in crowded, unsanitary conditions. Nearly every one of these 'aviaries' contained birds with psittacosis. The parrot import embargo was not going to safeguard people when local stock was so evidently affected.

Meyer wanted authorities to place an interstate ban on the movement of the birds. His appeal fell on deaf ears until September 1932, when Mary Borah fell ill.

The daughter of a former governor of Idaho, and wife of maverick Republican senator William Borah used to keep dozens of budgies, finches and canaries which she let fly unimpeded around the couple's Washington apartment. 'It's bad enough to take birds from their forest home without cooping them up in cages,' said Mary, surely one of the more singular Washington hostesses, in a 1928 interview with the *San Jose News*.

Petite and bird-like herself, she wore child-size shoes and clothes and was known as 'Little Borah'. But, for all the evidence that her husband was a dedicated womaniser who fathered a child with her friend Alice Roosevelt Longworth (eldest daughter of president Teddy) 'Big Borah' loved Mary fiercely and, after she became desperately ill, he convinced Hoover to also embargo the interstate movement of budgerigars and their broader family.

Little Borah was saved. In fact, she went on to outlive her husband by 36 years, dying at the age of 105, but in 1932 it was a close-run thing. It took no fewer than five injections of serum to put Little Borah right, obtained from the blood of two scientists who were themselves still recovering from laboratory infections they'd picked up in the 1929–30 outbreak.

Mrs. Borah is improved

Boise, Idaho—Mrs. William E. Borah, wife of the Senator, seriously ill from psittacosis, or parrot fever, received her fifth injection of serum today and was reported improving.

Her recovery received a brief setback late yesterday, but a bulletin from Dr. Ralph Falk early today said: 'Mrs. Borah has shown steady improvement since midnight'.

The first injection of serum, prepared from the blood of survivors from the disease and sent here by airplane from the laboratories of the Bureau of Public Health in Washington, was given at 4.30 am Saturday. Every twelve hours it has been administered since, and each bulletin, except that of last night has indicated steady improvement, strengthening of the heart and clearing up of the lung congestion which first led physicians to a diagnosis of influenza.

Dr Falk talked by telephone to Dr. James B. Luckie in Pasadena, Cal, who has treated nine cases of psittacosis. Dr Luckie promised to provide from his recovered patients all the serum which may be needed for Mrs. Borah.

—*The New York Times*, 27 September 1932

It was thought the birds were contaminated locally. No one then suspected budgerigars carried psittacosis in the wild in Australia. When Karl Meyer got permission to import 200 budgies from Australia, it was with the idea of perhaps inoculating or replacing local stock. But when, after quarantine, the imports began to die and tested positive for psittacosis, it became clear that budgerigars were every bit as responsible for spreading the infection as the parrots from South America.

In the four years after Mrs Lillian Martin was first diagnosed, there were 189 cases of psittacosis and fourteen suspected cases, with 40 deaths in the US and Canada. In 170 cases the afflicted person had had direct or indirect contact with diseased budgerigars. In the first seven months of 1934 there were 145 cases of psittacosis in Germany, resulting in 26 deaths. Of the birds examined in relation to these cases, 52 were found to be infected with psittacosis and 49 of them were budgies.

It was enough to attract the interest of one of the greatest minds of science, Frank Macfarlane Burnet, who determined that 'knowledge concerning the condition of native Australian parrots in the wild was urgently needed'.

The beetle-loving bank manager's son from Traralgon in Gippsland, Victoria, who won the Nobel Prize in 1960 for his discovery of acquired immunological tolerance, was soon deeply engaged with parrot spleens. Almost 300 lorikeets, cockatoos, rosellas, grass parrots, budgerigars and elegant parrots were killed with coal gas for the greater good.

Burnet found 'the widespread prevalence of psittacosis in Australia indicates that the enzootic conditions in budgerigar breeding establishments in Europe and America are probably derived from the natural infections of the original Australian birds from which they are descended'.

Further, he concluded that if the presence was confirmed in wild-caught birds in South America, 'we shall have to regard psittacosis as an almost universally present low-grade infection of parrots which only on rare occasions flares up into the dramatically infective human disease which characterised the outbreaks of 1929 and 1930'.

Meyer had already deduced the stressful, unhygienic conditions the birds were being reared in caused the bacterium to flare up and become virulent, killing host and handlers. Together with fellow scientist Bernice Eddie he instituted a program to help establish psittacosis-free aviaries, which involved testing 30,000 budgerigars. It was during this time that Meyer himself became infected with psittacosis after he removed his gloves to take a phone call. Fortunately, he did not become seriously ill.

The parrot bans around the world affected every fancier from king to commoner. King George V, for example, had acquired a mate for his macaw, but while it was en route from Brazil the prohibition was imposed in the UK. The king's private secretary, Clive Wigram, asked the health minister, Arthur Greenwood, for a workaround. This led to the extraordinary circumstance of Buckingham Palace being declared a zoological garden, taking advantage of a special clause permitting parrots to be imported by zoos.

The Norwegian-American stage and film actress Greta Nissen was not so fortunate. Caught between quarantine laws when their mistress travelled to the UK for the filming of the 1937 thriller *Café Colette*, Nissen's two blue budgies shuttled back and forth across the Atlantic three times before they were finally accepted at a French port.

In time scientists would recognise that, while *Chlamydia psittaci* was particularly common among parrots, birds of all types acted as a reservoir. Even mowing the lawn and running over some dried duck poo without a grass catcher could lead to disease transmission. Having been documented in 467 species in 30 bird orders around the world, psittacosis might be more fairly described as 'ornithosis' or 'avian chlamydiosis'—as it is also called.

Budgerigar

 Town warned 'escaped bird can kill you'

A police warning went out to Southampton last night: Dial 999 if you see an uncaged budgerigar. But don't touch it—may cause death.

Radio-car crews and men on the beat kept an eye on roofs, window ledges and parks for the bird. It is believed to be suffering psittacosis—parrot's disease which can kill humans.

A cage of nine budgerigars from South Africa was given to Hollybrook Children's Home, Southampton, by men of a liner's crew a year ago.

A few days ago one of the birds became ill. Nobody thought much about it. But yesterday a vet diagnosed the disease.

The Ministry of Agriculture and Fisheries ordered all the birds be destroyed.

Then it was discovered that one budgie was missing.

The hunt started. Police emphasised: Don't take risks if you find it. Dial 999 straight away.

And ten children at Hollybrook are being kept under strict observation.

An epidemic of parrot's disease killed 100 people in Britain in 1930. The import of all the parrot family birds, including budgerigars and cockatoos was banned. This was relaxed in January 1952 [1951].

But a couple of months earlier a Birmingham pet shop manager was killed by the disease. More people became ill. The ban was re-imposed again last February.

—*The People* (UK), 28 June 1953

Psittacosis is not the only means by which a budgie might occasionally kill a person. In 1958 virologists launched an investigation into bird-borne transmission of poliomyelitis after a budgerigar recovering from mysterious paralysis of its legs pecked the lip of a nine-year-old boy. Two weeks later the boy developed poliomyelitis and subsequently died, and the same poliomyelitis virus, type 1, was serologically isolated from both the boy and the bird. However, investigations concluded that birds were highly resistant to most strains of polio and it is more likely they very occasionally contracted the virus from people rather than the other way around.

In another case, reported in the *European Respiratory Journal*, a 50-year-old man was found to have died from 'chronic extrinsic allergic alveolitis', an inflammatory lung disorder now known as hypersensitivity pneumonitis. The cause of his severe allergic reaction was eventually traced to a wreath of budgie feathers that hung in the living room of the man's home.

In 2016 the same condition very nearly killed a 61-year-old retired nurse from Merseyside. She had bred budgies for nearly twenty years when she suddenly turned grey and was unable to breathe because of the cumulative effect of the plumes and dust.

While making for great headlines of the 'Man Bites Dog' variety, dying at the wings of a budgie is far more likely to be the result of human misadventure.

 Killed chasing budgerigar

Falling downstairs while chasing a budgerigar, Thomas Gallagher (47) of Milbourne Street, Blackpool was killed.

A verdict of accidental death was returned at the inquest last night.

Gallagher was with another man in the room in which the bird was flying about, it was stated. The bird flew through the doorway and when the men went to catch it Gallagher overbalanced and fell.

—*Dundee Evening Telegraph*, 16 August 1939

There have been few deaths in Australia due to psittacosis (or any other circumstance that might be pinned upon a creature weighing 35–40 grams or thereabouts), but it remains a notifiable disease, with 1687 cases reported nationally between 2001 and 2014.

A dose of antibiotics is usually enough to stop the disease in its tracks, but birds were never again transported so blithely across borders.

Warning to children who catch budgerigars

Children who have found a new way of earning money by catching lost budgerigars were warned yesterday that the birds might be carrying a disease which could kill them.

The warning comes from Dr Hilgrove Robinson, Medical Officer of Health for Carlton, Nottinghamshire where a child died within 48 hours of catching the disease psittacosis.

Dr Robinson says in his annual report published on the weekend that he believes the child caught the infection from a budgerigar.

Investigations in the neighbourhood revealed that several budgerigars were sick and a number of children had feverish colds. One boy was found to have psittacosis.

Dr Robinson spoke yesterday of the growing habit among children of catching escaped budgerigars which were prone to pick up the disease from sparrows, pigeons and seagulls. Some people paid children up to 10s for the return of budgerigars.

Dr Robinson said: 'I know one boy was recently trying to locate the owners of eight budgerigars which he had captured.'

He advised owners of recaptured birds to watch for any sign of illness. If a bird appeared to be sickening, any feathers it shed should be picked up with tweezers and taken to the health department.

—*The Birmingham Daily Post*, 30 January 1961

But while in some rare instances association with budgerigars might yet still kill, the little pet is more likely to be a saviour. In the 1960s and 1970s the Budgerigar Information Bureau—an ingenious public relations initiative by Pedigree Petfoods, which had the dollars to enlist the likes of Twiggy, one of the most famous models of the time, for a photo shoot at its London HQ—introduced bravery awards for the great Australian pocket parrot.

The very first inductee on the Budgerigar Roll of Honour was Joey, a ten-year-old bird who, in 1967, woke his 72-year-old sleeping owner Charles Davies when fire broke out in their Coventry flat by pecking him repeatedly on the face. Joey was awarded a gold-plated cage and stand together with a budgie bravery certificate. Blodwyn Cater saw her John similarly rewarded for screeching to wake her up when their shared home in Merthyr Tydfil in Wales caught fire in 1974.

Other acts of budgie heroism have been a little more ambiguous. Another Joey simply flew out of a house in Liverpool after a chip pan caught fire, whereupon one of the owners rushed back inside and discovered the blaze.

Budgerigar

The sudden death of a lamentably unidentified pet budgerigar caused Norfolk man Jack Ayres to investigate his immediate surrounds and discover gas leaking from a tin of cyanide in his pantry. It was just in the nick of time: he passed out on the street after he fled from the house.

Chapter 13

By royal command

IN COLONIAL TIMES nothing said 'I love you, Your Majesty' like a cache of cute critters brought with great difficulty from places afar. In July 1803 the French explorer and cartographer Nicolas Baudin turned the corvette *Géographe* towards home, abandoning the final survey work of his *Expédition aux terres australes* commissioned by Napoleon Bonaparte.

Originally built as a twenty-gun Serpent-class naval vessel, *Géographe* had had an extra deck added to accommodate men of science and the menagerie they collected. By the time the expedition headed west from Timor, some of the last surviving dwarf emu subspecies indigenous to Kangaroo and King islands had already been at sea for months.

They were not well—and nor was Baudin, who was spitting blood. On 3 July 1803, he wrote:

At daybreak we found two of our kangaroos dead in their pens. We could only attribute this to the violent and incessant movement of the heavy sea, which left them not a moment's peace. This news was particularly unpleasant, as I saw myself on the brink of losing them after giving them such attention as should have secured them a happier fate. [And] since the emus refused to eat, we fed them by force, opening their beaks and introducing pellets of rice mash into their stomachs. We gave them, and the sick kangaroos likewise, wine and sugar; and although I was very short of these same things for myself, I shall be very happy to have gone without them for their sake if they can help in restoring them to health.

Baudin never reached home but did at least make it to what was then French territory, the Isle de France (Mauritius), where he died of tuberculosis on 16 September 1803. Two of the surviving emus and many of those 50 other species of bird, including a pair of black swans, were finally delivered in late March 1804 to the soon-to-be Empress Joséphine, to stock the gardens of the Bonapartes' fabulous Château de Malmaison in Paris.

History does not tell us who was responsible for delivering Queen Victoria the rather more modest offering of a pair of budgies, but delivered they were in 1845—five years after John Gould returned to England with the first live pair. It was the beginning of the royal family's long association with the bird that would colonise kitchens and sunrooms of houses throughout the empire and beyond.

Aviculture in general, and parrots in particular, have long been affiliated with the ruling classes. As early as 1500 BC Queen Hatshepsut of Egypt was apparently collecting exotic birds from far beyond her realm. A bas-relief in her temple at Deir el-Bahari on the banks of the Nile depicts a secretary bird from Africa.

In the fourth century BC, the armies of Alexander the Great brought parrots back to Greece from the Punjab.

Roman nobles housed their birds in cages made of precious metals, and in the Renaissance, birdcages were commissioned works of art. As a teenage Tudor bride Catherine of Aragon bought a pet parrot with her from Spain. Queen Elizabeth I maintained a number of aviaries while King James I, and his son Charles II, kept a vast number of exotic birds along the road that runs down the southern side of St James's Park, London, and that is still known as Birdcage Walk. By the Victorian era, ornamental cages constructed in all manner of architectural style were regarded as essential fixtures in any genteel woman's parlour, and an aviary was a must-have adjunct to the greenhouse of many a stately home.

King George V's attitude towards fauna was typical of the early twentieth century. He professed to love birds but also had no compunction about personally shooting more than a thousand pheasants over six days in 1913. This same man was so indulgent of his pet African grey parrot Charlotte that he let her peck at lumps of sugar, the yolk of guests' boiled eggs and the odd morsel of bacon at the breakfast table.

He was also very partial to budgerigars and kept an aviary of them at Sandringham Estate. In 1930 the king accepted an invitation to become president of the world's first dedicated Budgerigar Club, established in 1925, which changed its name to the Budgerigar Society at his request. He was said to be delighted to receive a gift of 'royal blue' budgerigars that had been specially bred by well-known Sydney fancier Harold Pier. Presented to New South Wales governor Sir Philip Game on the king's behalf in March 1934, the royal budgies were eventually safely delivered unto the king's private secretary.

During one of his increasing bouts of illness, a pair of blue budgerigars kept the king company and became remarkably tame. On 20 January 1936 King George, a heavy smoker who suffered from chronic obstructive pulmonary disease, died aged 70. It was the monarch's beloved parrot Charlotte who accompanied his body on the final train journey from Sandringham to Westminster Hall for the lying-in-state.

 ## *Late king's parrot*

Unfounded report of its death

The late King's parrot Charlotte has experienced a distinction which is usually reserved for noted men. She has received a premature obituary notice. A report was published yesterday in a Sunday newspaper that the famous parrot was dead, that her carcass was in the hands of a taxidermist, and that she was destined to be exhibited in a natural history museum.

Charlotte was secured by King George when he was on active service in the East during his naval days, and she was his constant companion up to the time of his death.

When the Court moved from Buckingham Palace to Sandringham, Windsor, Balmoral, Charlotte always accompanied the King. Her last journey in the company of his late Majesty was when his body was brought from Sandringham to London in January last.

—*Western Morning News* (Plymouth), 29 June 1936

As a child, Princess Elizabeth used to delight in visiting the 5-square-metre budgie aviary at the Royal Lodge in Windsor Great Park. In 1946 she, her sister Margaret and their father King George VI were pictured standing in front of the budgie aviary.

A home movie made at the time, but not publicly aired until 2006, shows the teenager who would be queen inside the aviary laughing as budgies clamour all over her.

Two years before the death of her father and her ascension to the throne, Elizabeth exhibited twenty of her own budgerigars at the National Exhibition of cage birds at Olympia. It was the first and last time that the princess exhibited any birds under her own name and they were not entered in any competitive class.

 ## Princess trains Australian parrot to speak

Princess Elizabeth is training a young Australian budgerigar with the help of Her Majesty Queen Elizabeth. She has already taught it a number of words and is now trying to teach it to say a complete sentence. Princess Elizabeth is very patient with the bird, spending half an hour daily on its lessons.

The bird is one of 40 budgerigars and parakeets which are kept in an aviary built by Queen Alexandra at Buckingham Palace. The Princesses have read all they can about the birds and their habits and haunts in Australia.

—*The Advertiser* (Adelaide), 1 March 1938

The present Windsor aviary was designed by Princess Margaret's then-husband Lord Snowdon, as a test run perhaps for Britain's first public walk-through aviary which he conceived and co-designed for London Zoo in the early 1960s.

In 1970 the queen found common ground with Tito, Yugoslavia's communist revolutionary president, in the budgie aviary when he stayed at the royals' weekend retreat on a state visit. Tito became a huge fan of the little birds and had an aviary for them built on his private island off the coast of what is now Croatia. When the queen returned the favour and made her first-ever visit to a communist state in 1972 she took a dozen pairs of budgies as a gift for the 80-year-old dictator.

As children Prince Charles and Princess Anne were encouraged to keep the family connection with birds going by the Queen Mother, who gave them both blue budgerigars for Christmas in 1957.

Budgies are still very much part of the royal household's menagerie, with the queen's flock of 100 reported to have lost their zest for breeding in more recent years. Gardener and designated 'Keeper of the Royal Budgerigars' Graham Stone made a public appeal through the press in 2000 hoping to swap some excess males for new female birds. Naturally, as they would be mixing with royal birds, the newcomers needed to have the right pedigree.

'What I want to do is stop in-breeding in the flight so I am appealing for a female Liberty. We have about 80 or 90 budgies and each year the number goes down so I want to address the problem by introducing some new blood,' said Stone, the third official Keeper of Royal Budgies, without a twinge of irony. The queen's birds are free-flying, able to come and go through from the aviary via a wire mesh tube.

The training of the first homing or Liberty budgies is credited to Hastings William Sackville Russell, the twelfth Duke of Bedford. Known as 'Spinach' Tavistock (his previous title had been Marquess of Tavistock), as much for his reputedly wilted personality as his vegetarianism, he was a keen ornithologist but apparently a terrible father. His son recalled being so neglected he had to 'eat chocolates left out for the birds'.

Russell produced a book called *Parrots and Parrot-like Birds*, first published in 1929, in which he devoted more pages to budgies than any other species. He described the best methods of breeding the birds, training them to talk and, most remarkably, revealed their ability to home like pigeons:

> *As a pet, the Budgerigar possesses almost all the virtues of the larger parrots without any of their vices. It is lively, friendly, and amusing and when at liberty in the room, though they may occasionally nibble papers, it lacks the strength of beak to do any serious damage. By reason of its small size few of the noises it makes are loud enough to be irritating; while its capacity to learn to talk in a low, husky, but often perfectly clear tone is very great indeed.*

A second smaller booklet, *Homing Budgerigars: Their care and management*, followed in the 1950s, in which he enthused:

> *One of the most beautiful additions that a nature-lover can make to the amenities of a country garden is a flock of budgerigars trained to fly at liberty during the day time and return to the aviary to nest, roost and feed. The elegant form of the little birds, the almost endless variety of exquisite colours in which they are now bred, their swift and graceful flight, and their amusing ways provide an endless source of amusement and pleasure.*

Both the misanthropic duke's publications were rather popular—evidently the only thing about him that was—and before his sudden death in 1953 he was able to complete much of the second revision of the seminal work. The memoriam published in this book would have readers believe he died in defence of his budgies:

On October ninth, His Grace, who had previously travelled down to his home with some Budgerigars he planned to add to his group of homing Budgerigars, went out with his shotgun to hunt a Sparrowhawk which was menacing his flock of one hundred birds. While forcing his way through some bushes he seems to have stumbled and the gun accidentally fired, inflicting fatal wounds.

When his disappearance was reported commandos, police and volunteers scoured the moors of Devon. Commandos from a nearby Royal Marine school used walkie-talkie sets and mine detectors. In all two hundred men joined the search. Saturday night, October tenth, workers drained a large pond on the Duke's twelve-thousand-acre estate.

A gamekeeper's hunch finally led to the body. Estate workers found the Duke lying in undergrowth near a favourite beauty spot. Death was believed to have been instantaneous.

Both at Endsleigh and Woburn [family estates] it was the Duke's practice to release his homing Budgerigars early in the morning where they could fly at liberty in the garden and return to the aviary to roost, feed and breed.

His son, John Ian Robert Russell, who became the thirteenth Duke of Bedford, observed with considerable asperity that his father died in such a way as to create as many death duties as possible. The new Duke of Bedford inherited crumbling piles and huge tax debts, and became one of the first of a generation of gentry to turn his ancient family seat Woburn Abbey into a tourist attraction, complete with zoo and homing budgies. Accused of being undignified for taking part in game shows and inviting people to pay to dine with him, he agreed: 'I am. If you take your dignity to a pawnbroker, he won't give you much for it.'

It was in this spirit perhaps that Margaret Campbell, the former Duchess of Argyll, announced she was going into the tourist business, throwing open her family home in Upper Grosvenor Street, London. The society beauty, who had first been married to American millionaire Charles Sweeny and mentioned in this capacity in the Cole Porter hit 'You're the Top', found herself broke after her second marriage to the eleventh Duke of Argyll ended scandalously in 1963, with nude photographs and claims she had slept with 88 men.

 *Menagerie à trois*

Backed only by a budgerigar and two poodles, Margaret, former Duchess of Argyll, is going to challenge some of England's menagerie-equipped stately homes for tourist dollars, she announced yesterday. For $18 the main attraction at 48 Upper Grosvenor Square, an 18th-century house opposite the United States Embassy, will be champagne and the hostess herself.

—*The New York Times*, 10 April 1975

Chapter 14

The sealed section

AS THE STORY goes, the Duchess of Argyll had become sex-obsessed after plummeting down an empty lift well on a visit to her Bond Street chiropodist in 1943. 'I fell forty feet to the bottom of the lift shaft,' she wrote in her memoir *Forget Not*. 'The only thing that saved me was the lift cable, which broke my fall. I must have clutched at it, for it was later found that all my fingernails were torn off. I apparently fell on to my knees and cracked the back of my head against the wall.'

The brain injury caused her to lose all sense of smell and taste, and, it was said, any pretence to sexual monogamy. For a long time, the budgerigar laboured under a misapprehension of fidelity, much as the duchess had done.

In the wild budgies find their perfect partner and bond in long-term partnerships, exhibiting what ethologists call social monogamy. Among socially monogamous birds, the male takes an active role in parenting. He looks after his incubating mate, bringing food to her and to the nestlings as they hatch. But a DNA test might reveal the young the male tends so carefully are not his. For while our budgie couple do prefer one another's company and always attend parties together, they do not always keep their cloacas to themselves, as is demonstrated in the aviary environment—much to the frustration of serious breeders who are trying to play chess with genetics.

Interestingly, repeated studies have shown that male birds particularly are far more likely to indulge in 'extra-pair' relationships—a budgie on the side—when their mates are busy in their nest boxes and can't see them. The behaviour suggests, perhaps, that they inherently know what sides their nests are feathered on and don't want to get sprung. As the Duke of Bedford put it:

The morals of the cock budgerigar of the domesticated strain, unlike I think those of the wild bird, are it must be admitted, indeed conspicuous by their absence. A cock will pair with any hen who will give him the least encouragement, but he will only help in feeding the young of his first 'official' wife, his other lady friends being left to bring up their family single-handed. Although quarrelsome over their nests, hen budgerigars do not, as a rule, display much jealousy over their mates, whose morals they seem to regard as being beyond hope.

The duke observed that it was not uncommon for female birds to decide to share a mate and even, on very rare, occasions, a nesting box. The cock bird, he observed, 'obviously enjoys having two wives, but if they present him with twelve children, he will learn that after supper, comes the bill!'

It is with a certain amount of trepidation the pubescent budgie might google 'penis' and then discover it doesn't have one. Indeed, fewer than 3 per cent of birds do. Lesson 101 in budgie sex is that birds have only one orifice, the cloaca, which serves not only to excrete urine and digestive waste but also to discharge sperm or eggs depending on whether they are male or female.

When hormones are stimulated during mating season, the male begins to accumulate sperm in his cloaca, and the cloacal vent of both male and female starts to swell until it protrudes slightly. When the birds are ready to mate the male hops nimbly atop the female who, if receptive, obligingly moves her tail feathers to one side and raises her wings just a little in invitation as he gently 'treads her', draping his wing over her shoulder for balance rather than modesty. The birds then rub cloacas together in what is known as a cloacal kiss or, rather less romantically, cloacal apposition. By this means sperm is transferred to the female's cloaca, where it fertilises the eggs she is busily producing.

Eggs are formed one by one, and the hen will lay her first egg within two weeks of mating, producing an average clutch of four to eight eggs laid two days apart. Even before this the male budgie will dance attendance on the female, regurgitating food at the entrance of the nesting hollow where she has ensconced herself.

While the reproductive act itself is over in seconds, a lot goes into budgerigar courtship, some of which is much more subtle than the human eye can detect. If you want to know if your budgie is truly a stud muffin, hold him under an ultraviolet light. If he glows like he's wearing a white T-shirt at an 1980s disco, there's a pretty good chance the girls will go for him.

Budgerigars, male and female, have a naturally occurring yellow fluorescence in the feathers on the top of their heads and cheeks. These feathers absorb UV light, then re-emit it in dazzling neon. We can't see it with our poor feeble eyes, but birds can.

They see more of the colour spectrum, and there is mounting evidence that the more a budgie glows the more attractive it is to the opposite sex.

One of the more curious scientific experiments to help establish this was conducted at the University of Queensland in the Commonwealth-funded Vision, Touch & Hearing Research Centre (since succeeded by the Queensland Brain Institute's Sensory Neurobiology unit). The research, published in 2002, involved smearing the bioluminescent area of birds with sunscreen, which blocks UV light, or Vaseline, which doesn't. Given the choice, the budgies quite sensibly preferred birds not smeared with anything, but given no choice they went for those wearing petroleum jelly, which let their light shine through.

There are also other physical manifestations when a bird is getting ready to breed. The cere—the small, fleshy patch above the beak—changes colour. The male's cere turns darker blue and the female's goes dark brown and scaly. In wild budgerigars the blue/purple cheek patch is also of great significance.

'The female will look at that purple patch,' Dr Rob Marshall explains. 'It is a reflective patch so when a bird is in perfect condition and comes into what is called breeding condition—that is a heightened level of condition when the feathers become very bright—that purple patch will reflect light like a diamond, or that is probably how they would see it, as a very bright light just on the side of their beak where they perform to the female looking face on.'

In addition to these changes in birds' condition there is also the requisite amount of singing and dancing that goes into budgie courtship. 'Males will dance around a female. They will form a circle and they will dance in a circle and . . . it is quite weird, but I relate it to a corroboree type thing. It is a real dance. Five or six males will dance around the female performing to her—a bit like New Guinea bowerbirds or birds of paradise,' says Marshall who, in addition to working with budgies as an expert avian vet, is one of the few who have studied the birds extensively in the wild.

Sound and smarts are other key criteria in budgie mate selection. In addition to the usual feather fluffing, head bobbing and beak tapping, a hen bird is listening for just the right note in her would-be mate's song, and for his initial contact calls to be similar to her own, not just a mimic of her call. A 2006 study conducted at the University of California showed that male budgies whose female partners produced similar-sounding contact calls to their own were inclined to be more helpful around the nest.

More recently a collaborative study by scientists from China and the Netherlands showed that female budgies will actually change their mate choice if a less colourful or musically attractive budgie demonstrates more brains by problem-solving. The role of cognitive ability in sex selection offers fertile ground for further investigation by nerdy guys everywhere.

The male budgie's song or warble is also incredibly important in triggering females to ovulate and coordinate breeding. Studies have shown even females kept in isolation will lay eggs if a budgie warble is played to them.

All of this doesn't count for much, however, if you are stuck in a cage by yourself with only a mirror and a giant non-budgie for company ... which brings us to RBS, or Randy Bird/Budgie Syndrome. This is a topic that generally provokes snickers as people, even vets, do not always take bird sexuality seriously. James Allcock, one of the first vets to have a media profile—he wrote for *The Times* in England, and had a phone-in TV and radio spot—once famously canvassed a live call from a woman who was worried her budgie was masturbating on the bar of his cage. 'I can assure you he won't go blind,' Allcock replied, somewhat frivolously.

This is a serious and well-recognised condition, which Dr Peter Wilson of Currumbin Valley Veterinary Services, and founder of Australia's first dedicated bird, reptile and exotic practice, explains.

 ### Randy budgie syndrome

It is natural for a bird to reach puberty and choose a mate. It is unnatural for pet birds to be isolated from their own kind and restricted to a caged environment. Well-meaning owners often provide a mirror for company. This is the worst thing that they can do. The sexually frustrated, single pet bird will often try and 'bond' with his own reflection in a cage mirror. 'Randy Budgie Syndrome' is a recognised medical condition where a single, pet, male budgie endeavours to maintain a sexual relationship with his reflection. He masturbates on his perch or cage toys and regurgitates food to his reflection.

Some owners consider this activity as a form of entertainment, while others find it distressing. Such activity on a constant year-round basis can lead to digestive and hormonal disturbances. Frustrated, single pet birds will often engage in stereotypic and obsessive-compulsive behaviour. Some birds will continually pace up and down the length of their cage. Others will acquire a 'drinking problem'. This is a form of displacement activity where the frustrated bird channels its sexual urges into an obsessive-compulsive activity such as excessive drinking.

I always tell owners that they should never become a 'bird-ophile' in their relationship with their pet. There are appropriate and inappropriate ways of handling pet birds. In other words, owners should not touch or handle their pet bird in inappropriate or sexually suggestive ways. They should never allow the bird to eat out of their mouths or stroke it on the lower back or abdomen if the bird is presenting. These types of behaviours are 'birdy foreplay' and encourage sexual and mating behaviour in the bird.

The obvious way to counteract aberrant sexual behaviour in pet birds is to introduce a mate of the opposite sex. There are many 'old wives' tales' about having a mate for a pet bird. The most common misconception is that your bird won't be tame or talk if it has a mate. This erroneous idea has been disproved so many times. Instead of having one friendly little bird, you have two (provided recognised training and behaviour is applied). When birds have mates of the opposite sex, they have a natural outlet for their sexuality when they become sexually mature.

—Dr Peter Wilson, Currumbin Valley Veterinary Services; reproduced with permission

Paradoxically, the male budgie's deep desire for a mate goes to the very heart of its appeal to the human population says Professor Louis Lefebvre, a Canadian ornithologist and animal behaviourist based at McGill University, Montreal.

'Female budgies are horribly boring. It is the male budgie that is the success story, the Aussie male charmer that has made its way into the houses and hearts of millions of lonely people,' says Lefebvre. 'But male budgies are basically sexually frustrated constantly and they are forever trying to seduce and charm the people they are living with. The female budgie is just not that desperate.

'In this context you could see the Aussie term "mate" as a kind of global misunderstanding.'

Chapter 15

When duty calls

NOW PAIR-BONDED, as it were, for almost 100 years, both human and budgie endured many privations after Hitler's troops marched on Poland in 1939 and the world found itself once again at war.

While people on the home front were urged during both world wars to Save the Wheat and Help the Fleet, rationing and the controlled war economy would also affect what was in the tucker bowls of household pets in the UK.

Under the Waste of Food Order enacted on 12 August 1940 it became an offence to give pets 'excessive' amounts of food that might be considered fit for human consumption. While the Food Control Committee was advised that 'it was not the Minister of Food's intention to render the keeping of animals impossible when the use of human food for this purpose was reasonably necessary', it would come down to a matter of judgment whether that lump of meat for Fido warranted a summary conviction that might attract three months' gaol or a £100 fine—or both.

By then there were already considerably fewer pet mouths to feed as hundreds of thousands of panicked cat and dog owners had acted on widely publicised government advice and had their animals put down. The National Air Raid Precautions Animal Committee issued an Advice to Animal Owners notice in mid-1939, urging people to take household animals into the country in advance of any emergency. It concluded that if owners could not find a way of keeping their pets safe 'it really is kindest to have them destroyed'. It is estimated that as many as 750,000 cats and dogs were tearfully terminated within the first week of this declaration.

 ## Human food used for pet birds

Illegal and unpatriotic

A food ministry official referring today to the recent decision to stop the importation of seed for cage birds and pigeons, said that a number of complaints had been received from bird fanciers, some of whom had been foolish enough to say that if they could not get bird seed they were going to use other commodities such as oatmeal.

'This, of course, is illegal,' the official added. 'It is against the Ministry's orders to use human food in this way and it is extremely unpatriotic to use human food for feeding cage birds'.

The official pointed out that the total imports of feeding stuffs for cage birds if imported in the form of grain for fowls would be sufficient to feed 1,250,000 fowls which could produce about 400,000 eggs per day.

—*Dundee Evening Telegraph*, 4 March 1941

But a wee budgie could hardly hurt the wartime economy, could it? Certainly the editor of *Cage Birds* magazine didn't believe drastic action was necessary. 'I am anxious to counteract any idea that these pets have to be killed off. There is no need for it at all,' E.R.W. Lincoln told anyone who would listen.

Before the war Britain imported 300,000 tons of seed a year for caged birds, including 55,000 tons of millet, the budgie's favourite. But as the war went on the price of budgie seed mix went up from 5½ pence to as much as 2 shillings and 3 pence a pound, with the *Lancashire Daily Post* reporting queues at every pet and grain shop as word spread that supplies were close to exhaustion and further imports were to be banned.

It was not permitted to grow mere bird seed on vegetable plots, so some bird owners dug up their lawns to grow millet, linseed and sunflower. Others foraged for wild grasses, dandelions and chickweed and mixed in dried breadcrumbs, oatmeal and vegetable peelings to keep their birds nourished.

As bombs rained down on London and whole neighbourhoods were reduced to smoking craters, home-front propagandists evoked Britons' so-called 'Blitz spirit'. Even a budgie could do its bit to lift the national morale. 'Billy' the budgie seemed to get about a lot, often in unverifiable accounts, adding just a modest little chirrup of uplift.

 Child & bird saved from wrecked house

During the first raid in the London area bombs, which fell early in one district on the outskirts, completely demolished two houses, leaving a crater twenty feet deep. One of the houses was empty at the time. In the other house some of the occupants were killed. A boy, aged nine, who had been sleeping underneath the stairs, was rescued unhurt after being trapped for one and a half hours on the edge of the crater. While being rescued the boy showed no anxiety except for his budgerigar. It was his parents, as well as a woman who lived with them, who were killed.

—*The Manchester Guardian*, 31 August 1940

In addition to the usual repertoire of nursery rhymes and everyday phrases, many a bird was taught patriotic slogans, interjecting with 'Down with Hitler!' or 'Shoot Göring!' as neighbours swapped war news over a cuppa. A forward-thinking Liverpool woman trained her bird to quiz her, 'Got your gas mask, Mum?' when she was about to leave the house.

Newspapers were sprinkled with amusing little snippets of budgerigar derring-do. Take the family that returned after an air raid to find their dwelling gutted and no sign of the budgie. Days later they received word that the budgerigar had flown into a house in an adjacent town, uttered its name and address and concluded with the words, 'Naughty Mama, come and fetch me!'

But none were as redoubtable as Edith Binge and her budgie, whose 'Blitz spirit' was captured in a 'letter' first published in the London *Daily Express*. It was addressed to William Hickey, the pen-name of one of the paper's columnists—at the time it was journalist, MP and later suspected Soviet spy Tom Driberg who was concealed behind the Hickey name. The letter was widely republished in American, Canadian and Australian newspapers.

Edith Binge & the stovetop budgie

I had a seven roomed house. I have lived here (London) thirty years. I have been terribly bombed three times, from a seven-bedroom house now all bombed down I have got one room. I had a beautiful home and garden, 45 standard rose trees, etc not one left under the debris all uprooted and blown away.

I also had a very nice aviary with Budgys in, my love and pride they were all blown to bits. There was also six dear little eggs, one I found among the debris. I put it in an electric stove at a certain temperature and thank God it hatched and I have brought it up. I fed it from my mouth and it is the most beautiful blue.

All doors, all windows, all rooms upstairs are down. I am doing the repairs meself. I've put up bed-sitting room ceiling and enamelled walls etc, cut the six steps from the stairs in passage and filled in the space where the upstairs used to be. Today I'm using a pick and digging up the garden.

I've put a few things in the garden—a bush rose tree, tomatoes, potatoes and tulips, etc. I've had to put up me own fence so that I'm a little private. I live here all alone with my little bird. My son is in the army.

You see, sir, my bird and I live in a world all of our own. The gentlemen at the town hall say I'm genius, but I don't think so. Just comes into me 'ead and I do it and love every minute of it 'cause the harder I work the nicer the place is becoming. I'm now building a small summer house just for my armchair and my little bird so that later on we shall watch the roses bloom.

I'm only a little old lady. My soldier son calls me his little sixpenny worth of coppers. We love one another so much. I'm always in, working. I don't want to go until—if God spares me— until my son comes home and to the grand victory. Please excuse my bad spelling. You see, I'm 64 years young. I must close now, sir, God bless you and keep you safe.

—Edith Binge, 1942

(Left) Nineteenth century German ornithologist Dr Karl Russ was the bomb on birds, and parrots in particular. (Right) John Gould was credited with describing almost half the native birds of Australia and also relished eating them. Studio portrait cira 1860.

Artist Elizabeth Gould gave flight to her husband's soaring ambitions only to die shortly after giving birth to their eighth child. This image is from an oil portrait by an unknown artist painted after her death at age 37 in 1841.

Budgerigars were sentimental favourites of Elizabeth Gould who delighted in sketching them in the wild. *Melopsittacus undulatus* was one of 84 key plates she illustrated for *The Birds of Australia* (1840–48).

Before Athena Starwoman, budgerigars were the best-known Australian astrologers. This photograph of a fortune teller and her budgie was taken on the streets of London circa 1900. (Paul Popper/Popperfoto via Getty Images/Getty Images)

The world's biggest little movie star Shirley Temple is paired with the biggest little pet for this magazine cover promoting the 1939 film *The Little Princess*.

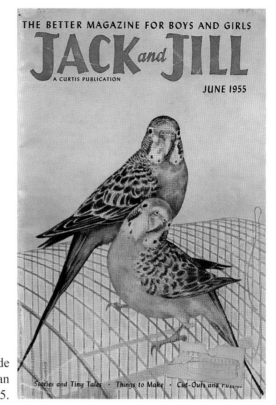

Wholesome and chirpy parakeets made the perfect cover for a popular American children's magazine in June 1955.

Never mind smoke in their eyes, cute finger-trained budgies were used to promote all manner of products including in this 1954 advertisement for Old Gold cigarettes.

How do you operate on a budgie? Very carefully. (Hulton-Deutsch Collection/ Corbis via Getty Images)

Peter the budgie went loco for his nine-year-old owner's electric train set in 1953. (Keystone Press/Alamy Stock Photo)

Although Sparkie Williams, the world's most famous talking bird, has been well and truly stuffed since he died in 1962, he still draws crowds to the Natural History Museum of Northumbria where he is permanently perched. Sparkie is seen here with his owner Mattie Williams in 1990. (Newcastle *Evening Chronicle*)

Sparkie Williams generated a lot of parrotphernalia. (Natural History Museum of Northumbria)

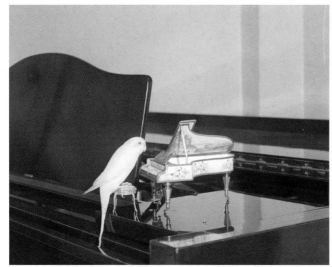

Chirpy, chirpy, cheep, cheep, a budgie tinkles the ivories of its own mini baby grand piano. (Ron Gerelli/Express/Getty Images)

In the 1960s Don Crown was a familiar sight on London's Southbank busking with twelve budgerigars and a dog. Sadly, his psychedelic first single 'The Budgerigar Man' (1970) and follow-up 'Mrs Wilson's Budgie' (1971) didn't cut through. (Trinity/Mirrorpix/Alamy Stock Photo)

The Good, the Bad and the budgie. Clint Eastwood smoulders with pet birds on his shoulder at home with first wife Maggie Johnson in 1959. (CBS Photo Archive/Getty Images)

Parakeets Bluebell and Maybelle were among the First Family's fondest pets during the presidency of John F. Kennedy. (Cecil Stoughton, White House Photographs, John F. Kennedy Presidential Library and Museum)

Parakeets proved popular props for promotional photographs. American actress and opera singer Kathryn Grayson is seen here in typical studio pose circa 1950. (Pictorial Press Ltd/Alamy Stock Photo)

Cast as Altaira Morbius in the 1956 sci-fi film *Forbidden Planet*, actress Anne Francis looked out of this world in publicity photos sporting budgerigar earrings and a skin-tight space suit. (Pictorial Press Ltd/Alamy Stock Photo)

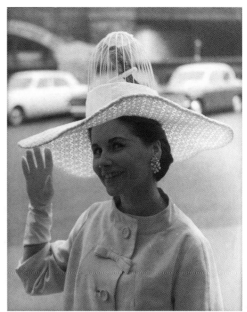

(Left) Double take, is that you Audrey? The unnamed woman with red hat, red lipstick and budgie bears an uncanny resemblance. (Popperfoto via Getty Images/Getty Images) (Right) What better to wear to the Birdcage enclosure at Royal Ascot than a birdcage hat complete with live budgie as sported by Jane Bough in 1968. (Keystone Press/Alamy Stock Photo)

Aussie, Aussie, cossie oi, oi, oi. The trademarked Budgy Smuggler swimwear line. (Stephen Dwyer/Alamy Stock Photo)

The famous Bird Lady of Cairo, Madame Amal Safwat, pictured with feathered friends on her lounge in 1999. (Photo Raphael Gaillarde/Gamma-Rapho via Getty Images)

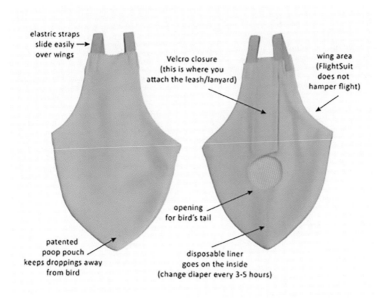

elastic straps slide easily over wings

Velcro closure (this is where you attach the leash/lanyard)

wing area (FlightSuit does not hamper flight)

opening for bird's tail

patented poop pouch keeps droppings away from bird

disposable liner goes on the inside (change diaper every 3-5 hours)

Popping on a FlightSuit prevents pooping on the furniture when budgie is let out to stretch its wings. (Courtesy of Avian Fashion)

(Left) And when you have a pair of dead budgies in the freezer, voila! The WoW Budgerigar Bra by Emily Valentine (Bullock), Australia. (Courtesy of World of WearableArt® Ltd) (Right) During the murderous millinery period millions of birds were killed to adorn hats. This silk bonnet circa 1890 featuring several dead budgies was a gift of Susan Dwight Bliss, 1937. (Metropolitan Museum of Art)

Aviary casualties are reborn as *Chirpy Chiwhawhas*, a work of budgie feathers and mixed media by Emily Valentine, 2006. (Courtesy of Emily Valentine/ Bullock)

Who's a pretty bird then? Social media star Cooper the Budgie tweets a strong ethical bird-keeping message to the world. (Courtesy of Michelle McKee)

One lost budgie survived a Winnipeg winter hanging out with house sparrows. MacGyver was saved by Shelley and Val Corvino who built a trap to catch him. (Courtesy of Shelley Corvino)

Splashdown! Wild budgerigars drop in for a drink at a waterhole in Karratha, Western Australia. (Photo Jim Bendon, Wikimedia Commons)

The look of love. A pair of wild budgies at a nesting hole. (Courtesy of Barney Enders)

Champion Victorian breeder and
ANBC Hall of Famer Alan Rowe ruffles
nary a feather among his prize flock.
(Photo Sarah Harris)

South Australian breeder
Kelwyn Kakoschke's
budgies are parrot pin-ups.
(Photo Sarah Harris)

Geoff Capes, twice the World's
Strongest Man, has always had a
soft spot for budgies. (Photo Tony
Evans/Getty Images)

Budgies, like many other pets, also served as a helpful early warning of air raids. As Brian Davey recalled in one of the BBC's *World War II People's War* stories: 'Just about five minutes before an air raid, the budgie would get agitated and ring his bell and get his head under the bell continuously,' Davey recalled. 'He must have felt the vibrations through his feet like all creatures do when thunder is coming, only it was German aircraft. Then five minutes later the air-raid warning would go off and we would run to the Anderson Shelter picking the budgie cage up on the way.'

 ## Patriotic budgerigar

A budgerigar which is known by the whole of the village of Rhuddlan as 'Billy' flies about the Marsh Hotel, and if it hears Hitler's name being mentioned it immediately flies into a rage, screeching 'Nasty man, nasty man. Three cheers for Churchill'.

'Billy is terribly patriotic,' Mrs Thomas Lee, the licensee, told a reporter. 'He has been able to speak since only a few weeks old, and now he is nearly two years old.'

—*Liverpool Daily Post*, 8 April 1940

 ## The 'bird' for Adolf

A woman in a north-west England town who took her pet budgerigar 'Billy' with her into an air-raid shelter today said afterwards: 'While we were in the shelter and during the morning he has been saying "Here's those planes again. Down with Hitler!"'

—*Northern Daily Mail* (Hartlepool), 28 August 1940

Budgerigar

 *The whistling budgie*

[W]hen demolition squads went to one of the houses they heard a cheerful whistling among the ruins. After lifting pieces of masonry and fallen beams they found a budgerigar singing happily in its cage. The wires of the cage were twisted and the bottom of the case was filled with splintered glass, but the bird was none the worse. Rescue workers are searching for any other victims.

—*Belfast Telegraph*, 16 August 1940

 Rescuing the 'budger'

[A] block of workpeople's flats had been bombed and was unsafe to enter. A man hanging around outside said, 'I have a little budger in there. I do want it and they won't let me in.' He almost held his breath when he crept in behind me. We got his budgerigar and three others out of those buildings, as well as fourteen dogs, four cats and five cages of canaries.

—*The Observer* (London), 18 November 1940

Before the war, the number of bird-keepers and breeders with aviaries (as distinct from those with birds as household pets) in Britain was put at 560,000. By the time the guns fell silent this number had shrunk to an estimated 140,000.

The chronic shortage of birdseed led to fanciers of all feathers coming together to form a peak body, the National Council for Aviculture (NCA), to better make representations to the food ministry. During the fanciers' meeting held at the Waldorf Hotel a flying bomb made a direct hit on nearby Smithfield meat market, rather underscoring the hopelessness of their case.

Animals in bombed areas

The work of rescue: canaries and budgerigars

The ambulances and inspectors of Ulster S.P.C.A. continue their work of rescue and feeding of dogs and other small animals and birds in bombed areas, including the country districts.

Many dogs have been got out of the debris and returned to their owners. As the ambulance patrols the Belfast district owners are bringing their canaries, budgerigars, dogs and cats and imploring the officer to put them out of their misery and terror.

Many evacuees have taken their animals with them, while those whose homes are still intact are generously keeping and caring for their distressed neighbours' dogs.

The society has endeavoured by every possible means to save as many animals as possible during the past week and begs owners to arrange temporarily for their keep.

—*Belfast Telegraph*, 23 April 1941

 ## 8000 animals rescued from flying bombs

Some of the busiest Rescue Units in London and Southern England are those of the People's Dispensary for Sick Animals. Their vans have rescued 8000 animals since the flying bomb appeared and more than 250,000 since the war began.

A six-wheeled rescue van came in with an odd collection—two parrots, two budgerigars, a canary, 14 dogs and several crates of cats. Some of them kept their mouths open because of shock. As soon as the van was emptied helpers began to make it ready for another mission.

—*The Manchester Guardian*, 3 September 1944

127

The situation was even more dire on the Continent. 'The art of aviculture was never practised in Europe on a larger scale or with greater success than between the wars,' our aforementioned French friend Jean Delacour lamented knowingly. After the whole of his collection and his family estate at Villers-Bretonneux had been obliterated during World War I, Delacour had re-established himself at Clères in Normandy, in a glorious sixteenth-century chateau. Slowly and thoughtfully he had rebuilt his collection of birds and animals through the 1920s, and it included a sizeable aviary of budgies of blue, yellow, apple-green and olive.

He managed to escape the estate after it was first bombed by the Luftwaffe on 7 June 1940 and made his way to America where he remained for the rest of the war. Over the next five years the park was hit repeatedly by bombs from both the Allies and Germans, leaving craters up to 16 metres across.

Yet Delacour's friends managed somehow to save 'a nice lot of animals and birds' from their destruction until the very last day of German occupation. 'It was a very bad stroke of luck,' Delacour wrote. 'SS troops, who had been prevented from crossing the Seine for several days by Allied bombing, managed to pass the river one foggy morning, and they went straight to Clères. Partly hunger and partly spite, they shot with tommy-guns everything alive they could find, just before they left for good.' Incredibly, Delacour would rebuild a third time and bequeathed Clères to the French people upon his death at the impressive age of 95.

Another gentle, brilliant bird man would not be so lucky. Delacour worked closely with one of the foremost conservationists of the day, an honourable life member of the Wildlife Preservation Society of Australia, Jean-Marie Derscheid. Like Delacour, the Belgian ornithologist had begun building his bird collection, including the now ubiquitous budgie, as a teenager on the family estate in Sterrebeek, north-east of Brussels.

Having served in World War I, Derscheid rejoined his army medical corps in World War II. When his unit was demobilised upon Belgium's surrender to Germany in 1940, he returned home to find that his bird behaviour research station Armendy Farm, and his estate, had been devastated by invaders. Derscheid joined the Belgian Resistance, helping many Allied soldiers and airmen escape until his arrest by the Germans in October 1941. He was imprisoned in Germany for two and a half years before being decapitated on the orders of Heinrich Himmler.

The budgies in Australia may have been nowhere near the horrific frontline action but they did their bit. Billie Peach (mentioned earlier) from Darling Point, Sydney, was enrolled as an honorary Red Cross Voluntary Aide. Exhibited at Taronga Zoo with a collection box for a week in September 1940, Billy raised 40 times his own weight in coins. By then he already had an armoury of patriotic phrases, including 'What are you doing pet? Making socks for soldiers' and 'God bless the AIF and bring them home safely'.

In Woy Woy on the Central Coast of New South Wales there was budgie Peter Casey, who earned extra credits because he belonged to Joe Casey, an ex-soldier who had been 'so severely injured during the Great War that he must spend most of his life now in bed', and Joe's wife Ruby. For three-quarters of an hour, *The Daily Telegraph* breathlessly reported, Peter was the 'sole entertainer' at the local Red Cross Association branch with a repertoire that included 'Where the Dog Sits on the Tuckerbox' and 'Mademoiselle from Armentières'.

Budgerigar

Budgerigar lights gas: owner fined

A budgerigar blamed for a black-out offence at Sutton Coldfield, was stated to have turned on a gaslight by perching on the chain . . . its owner was fined £1.

At 12.15am a special constable saw an upstairs window flooded with light at 30 Victoria Rd, Sutton. The occupant of the flat, Donald Herbert Harper, was not at home.

This letter of explanation was read to the magistrates:—

'Mr Harper has a pet budgerigar which is usually allowed out of its cage for half an hour each evening.

'A few hours previous to the light being discovered the bird had returned to its cage but the door had not been closed.

'While flying around the bird must have settled on the chain of the bracket and turned on the gas which is lighted by a pilot-jet.'

A police officer said the weight of the budgerigar was about three ounces and that would be sufficient to operate the bar-tap on which the chains hung.

—*Evening Despatch* (Birmingham), 30 July 1941

And, when the war finally ended, naturally budgerigars inveigled their way into the Potsdam Conference with the 'Big Three' Allied leaders. Surprisingly, though, it was not the animal-loving Churchill who brought 'love birds' to the historic court of Prussian kings—he didn't acquire his famous budgerigar Toby until 1954—but the Soviet dictator Joseph Stalin.

Two budgerigars, said to be Stalin's constant companions, were installed in a large gilt cage in his personal suite. The man held responsible for the deaths of millions of his countrymen reportedly uncovered and talked to his birds each morning as orderlies waited, napkins on arm, to serve him breakfast.

130

Chapter 16

The cult of the budgie

TWO EPIC WARS, the Great Depression, implication in a nasty disease transmissible to humans and import bans may have slowed the budgerigar's ascent to its position among the world's top pets, but these crises couldn't stop it.

The lifting of restrictions on the import of birdseed into the UK in May 1950 was followed by the scrapping of the 21-year parrot import embargo in 1951, which enabled the budgie to really take flight among fanciers.

When budgerigars very first joined the ranks of show birds at Crystal Palace in London in 1924, there was just a handful of entries for one class. By the time of the 1956 National Exhibition of cage birds there were more than 3500 budgies entered across dozens of categories, with the ranks of committed fanciers including everyone from Queen Elizabeth and Sir Winston Churchill to pensioners from council estates and short-sighted Liverpudlian lads like John Lennon. Imagine!

It was with good reason William Watmough titled his definitive guide to keeping, breeding and showing these birds *The Cult of the Budgerigar* (1936). 'What is the reason for this phenomenal development and this never-waning popularity?' the unflappable Yorkshireman asked, before plunging on to answer:

> *In the first place, and this is the keynote of the whole story, we have in the budgerigar a beautiful bird, produced in a comparatively large number of colours immensely pleasing to the eye. It is a creature of an attractive shape, with nothing of the grotesque about its form, a bird of curves and symmetry, of balance and charm. It is clean in appearance and habits, it is alert and graceful, always on the move—a bird of swift flight. Its twinkling watchful eye reflects its intelligence. Its merry ways are a constant source of delight.*
>
> *Budgerigars are easy to manage and at normal times cheap to feed. They are free breeders and, as a rule, wonderful parents. They are healthy and vigorous. They thrive exceedingly in almost any climate. The biting winds of winter in the far North of Scotland, the sub-tropical heat of summer in the South, hail, rain, frost or snow, do not disturb the equanimity of these feathered aristocrats.*

Watmough was an archdeacon of the budgie cult, chapters of which had spread throughout the world since the formation of the first UK budgerigar club. The club's membership was not made up of Bob-and-Beryl budgie owners of Bognor, but serious aviculturists: no fewer than seven fellows of the Zoological Society were included among its founding eighteen members.

At its peak in the mid-1950s the Budgerigar Society, as it was renamed, boasted 22,000 paid-up members. In 1954 it hosted the first-ever World Budgerigar Convention at the Royal Hall in Harrogate, Yorkshire, an event that was evidently a lot less funny than it sounded.

 British supremacy in aviculture

HARROGATE, Monday: The main roads leading here are emblazoned with bold traffic signs bearing the designation 'Budgerigar Convention,' an appellation which evokes in the untutored a vision of a multitude of green and blue birds happily chanting nursery rhymes in unison.

Such a vision is, however, far from reality, for the 330 avian delegates to the First World Budgerigar Convention which opens here tomorrow, are the nobility of the species, the well-bred aristocrats who shun the repetitive lispings of the plebeian parlour birds and talk only a language of their own.

—*The Manchester Guardian*, 17 August 1954

Millions more fanciers belonged to dedicated clubs and hobby groups that sprang up in almost every town across the UK. In the mid-1950s budgerigars were reckoned to be the most popular household pet in the UK, accounting for some 80 per cent of the estimated 6 million birds chattering in sunrooms across the country. At this point there were at least two large budgie breeding farms in Britain producing between them 50,000 birds annually, in addition to the many thousands of small-scale backyard breeders selling through clubs, classifieds, local pet shops and the like.

According to figures put out by the British policy think tank Political and Economic Planning (PEP), the cage bird population multiplied tenfold in the decade from 1947, putting the feathered pet population significantly ahead of 3.7 million dogs and 3.2 million cats. By this calculation there was a budgerigar in every fifth home in Britain, but at this time there was also a distinct lack of household statistical data available for the UK, with the 1931 census data destroyed by fire, the 1941 census cancelled due to war and the 1951 census omitting to ask any questions about pets. The figures quoted by PEP were instead drawn from one of the earliest exercises in market research, a non-compulsory questionnaire called the Hulton Readership Survey, which began with the query 'What newspaper do you read?'

A better measure of the budgie's popularity is how embedded it became in Britain's popular culture. Basil Thomas, a promising playwright whose career was cut short by his death at 44, made budgies the foil in his 1935 work *The Lovebirds*. The farce revolves around a widower and widow embarking on married life together only to discover the budgerigar they were given as a wedding present is possessed by the spirit of a jealous deceased spouse. (The play, incidentally, was adapted into the 1961 British comedy film *The Night We Got the Bird* in which the budgerigar is replaced in typical big-studio style with a showier South American parrot.)

But it was when a budgie named Joey joined the radio soap *The Archers* on BBC Radio 4 in 1953 that the bird really arrived in earnest as a key player. Joey, a bird owned by Bert Dibb from Birmingham, entered the storyline when he was given to Mrs Perkins by Walter Gabriel for Christmas, whereupon he dutifully uttered the line 'Happy Christmas'. It was an acting debut not to be sneezed at. The serial was at the time broadcast to 9 million listeners, reached 20 million at its peak, and was also transmitted in Australia, Canada and New Zealand. It continues today as the world's longest-running drama, with storylines reflecting current events.

The 'appearance' of Joey on the wireless inevitably caused a spike in the number of birds named for the radio star and no end of confusion when any got lost, with at least half a dozen budgies answering to the name of Joey reported missing in papers across the country on most days. A surprising number of these feathered expeditioners talked their way home by providing the finder with their name and address, and some had quite catastrophic encounters while at large. In one instance it was not the budgie that was lost as much as the owner.

 ## Mrs Beadle and Joey are reunited

A budgerigar named Joey, which had escaped from its home in Lewes four days earlier, was taken from a cat's mouth at Chiddingly on Friday—uninjured. He had flown 10 miles.

Joey's owner, Mrs A.W. Beadle, of 37 Evelyn Road, Lewes, had inserted an announcement about his disappearance in the *Express-Herald* lost and found column that day and it was through this she was able to recover the bird.

A friend of the cat's owner saw the announcement, and contacted Mrs Beadle by phone within a few hours of the *Express-Herald* being published.

The owner of the cat, Mrs E. Cosstich, of East Haven Cottage, Whitesmith, Chiddingly, was indoors on Friday when she heard a bird screeching. She went to the door and found her cat with Joey in its jaws. She took him from the cat, and carried him indoors, placing him on a table, where he revived almost immediately.

Mrs Beadle took Joey home the following morning, after identifying him by a number ring on his leg. Back in his cage, apparently uninjured, Joey spent the rest of the day and all Sunday sleeping off the experience.

—*Sussex Agricultural Express*, 4 July 1958

 ## Budgerigar talked his way home

Mr George Anderson, of Axwell Farm Whickham, County Durham, picked up an exhausted budgerigar in a field, took it home and placed it near the fire. When the bird recovered it said: 'I'm Bobby Ritson and I live on Whickham Bank.' Mr Anderson made inquiries and found the bird was right.

Bobby Ritson had flown out of the house on Eleanor Terrace, Whickham, where his owner Mrs Jane Ritson, aged 68, lives. After being missing for two weeks, he was given up as lost, but Mrs Ritson now says: 'I'm not worried if he does get out now. He can talk his way home.'

—*The Manchester Guardian*, 14 May 1954

 ## Lost memory woman with a budgerigar

A woman suffering from loss of memory is in the care of the staff at Guest Hospital, Dudley, with a budgerigar in a cage by her bedside. The woman, aged between 35 and 40, was carrying the bird in its cage when she was found wandering in the centre of Dudley on Saturday night.

The police report that she got off a Midland Red bus at Fisher Street bus station at about 8.30pm on Saturday. There was a possibility, they said, that she lived or worked in the Dudley Port area of Tipton, because in her blue checked belted coat there was a threepenny bus ticket issued by a Birmingham city transport vehicle. Dudley is served by the Birmingham transport service on the No 74 route which passes through Dudley Port.

The woman is described as 5ft 6in tall and of slim build. She has dark hair, which is greying, brown eyes and good natural teeth. There is a scar on her right wrist. Her clothing includes a pink cardigan, a brown skirt and gloves, a Fair Isles jumper, and Wellington boots. There were two packets of bird seed in her black plastic shopping bag.

—*Birmingham Daily Post*, 20 January 1958

Tony Hancock, one of the biggest names in British television after his comedy series *Hancock's Half Hour* made the successful transition from radio, famously dressed up as a budgerigar in a cage that horrified his spinster owner by reciting a risqué verse during a visit from the vicar.

'The Budgie Sketch', written by Ray Galton and Alan Simpson for the BBC's *Christmas Night with the Stars 1958*, became a Hancock perennial, outlasting the highly-strung star who committed suicide in a flat in Sydney's Bellevue Hill a decade later. The great Spike Milligan, who was fond of inserting a budgerigar reference or two himself in *Goon Show* scripts, noted of Hancock, 'He was a very difficult man to get on with. He ended up on his own. I thought, he's got rid of everybody else and he's going to get rid of himself and he did.'

From *The Goon Show* to *The Two Ronnies* . . . the little one of the partnership, Ronnie Corbett, joking about his own stature: 'I was cleaning out the budgie cage and the door slammed shut on me.' Boom boom! Tommy Cooper, another of Britain's much-loved comics, also sometimes added a stuffed budgie to the top of his trademark fez for a gag involving a lost pet and an empty cage.

The wee birds have played a small but dynamic role in some of the world's best-loved sitcoms including *Porridge*, where prison kingpin Harry 'Grouty' Grout (played by Peter Vaughan) ran operations from a luxury cell complete with a budgie named Seymour.

The fabulous Sid James as Sid Plummer consults his psychic budgie Joey who refuses to talk, but tweets when he hears a horse name that's a winner in the classic *Carry On at Your Convenience* (1971). James also shared a set with a budgie in the hit ITV sitcom *Bless This House*. This familiarity apparently qualified him for his 1974 role in a film called *The World of Budgerigars* sponsored by Pedigree Petfoods and the Budgerigar Information Bureau, just two years before he dropped dead of a heart attack on stage

Mrs Slocombe in *Are You Being Served*? played by Mollie Sugden, had a dimorphous budgie/canary called Winston, but little was heard of it on account of the popularity of her pussy. On *Coronation Street* Mavis Wilton's budgie Harriet dies, apparently due to the stress of moving house, only to be replaced by Harry and later joined by a budgie called Beauty, while *George and Mildred* have a little chap called Oscar, who also dies, underscoring the budgie as an important dramatic/comedic device in the absence of other killable characters.

Budgerigars also make excellent ensemble players. Budgie circuses, for example, have been a thing since the 1820s when Jean Pierre

Ginnett—a French cavalryman captured during the Battle of Waterloo and shipped to England—married a local lass upon his release and went into show business. His performances with a budgerigar and a pony were the foundation of the Circus Ginnett, which became one of the UK's largest circuses in England in the late nineteenth and early twentieth centuries.

In the 1930s, performing 'lovebirds' and their handlers could be seen busking at the end of the pier in popular holiday spots, and were also readily available for hire for parties and community events. Piper's Performing Lovebirds, for example, who appeared at the 1932 annual dinner of Cheltenham Cricket Challenge Cup Association, were hailed as 'marvels of sagacity, bowing, climbing rope ladders, and riding see-saws and miniature horses at the bidding of their trainer with the orderliness of drilled soldiers'.

Norman Barrett MBE, holder of the Guinness World Record for longest-serving ringmaster, is the world's best-known budgie wrangler—famous for his troupe of acrobatic birds. Barrett developed his bird act after he was forced by knee injuries to give up more athletic circus arts. He quickly learned to use only male budgerigars to prevent jealousies and fighting, and always has two sets of budgies fully trained so he can rest them. Fifty years on Barrett's budgerigars, variously named Pepe, Maurice, Harold, Klaus, Jean-Pierre, Freddie Halfpenny, Cyril and Edward, are still delighting audiences as part of Zippos Circus.

Seguing from budgies on bikes to Budgie bikes . . . in the early 1970s Raleigh's Chopper bicycle, based on modified or 'chopped' motorcycles and featuring in the 1969 cult film *Easy Rider*, became the must-have item for cool kids on the block. A version called the

Budgie that was rolled out for younger children also proved popular, until the advent of the BMX put Choppers in the shade.

In 1971 came the little bird's first TV title role with *Budgie* the television series. Former pop star Adam Faith starred as 'chirpy cockney petty criminal' Ronald 'Budgie' Bird.

Peter Edward Clarke, the brilliant long-time drummer of Siouxsie and the Banshees, is known professionally as Budgie, supposedly for his habit of wearing a black and yellow jumper with undulating stripes. The cult English punk rock band The Clash immortalised the bird with the line 'News flash—Vacuum cleaner sucks up budgie' in their 1981 single 'The Magnificent Seven' from the album *Sandinista!* Before this the budgie was better known musically as the name of one of the earliest heavy metal bands. Formed in Cardiff, Wales, in 1967, the group performed initially as Hills Contemporary Grass or Six Ton Budgie, then morphed into Budgie in 1968. Vocalist and bassist Burke Shelley remarked that he 'loved the idea of playing noisy, heavy rock, but calling ourselves after something diametrically opposed to that'. Budgie's eponymous first album was followed by *Squawk* in 1972 and continued the birdy theme with *Impeckable*, *Nightflight* and the most recent offering *You're All Living in Cuckooland* (2006).

Sarah Ferguson, Duchess of York, replaced the birds' wings with blades to whip up publicity with her children's book series based on the character Budgie the Little Helicopter. 'Budgie' was the name she had gave her first training helicopter, and she went on to get her helicopter pilot's wings, fulfilling a bet she'd had with Prince Andrew.

More recently budgies have popped up in *Harry Potter*, and hardcore fans know that when German Seeker Thornsten Pfeffer attempted—and failed—to perform the Wronski Feint during the first round of the 2014 Quidditch World Cup, he briefly believed himself to be a budgie called Klaus after a nasty knock to the head.

Even though now the bird's popularity has waned, it still holds a place in contemporary culture as top cage bird. You could still buy a budgie and all the accessories its little heart desired in the Bird Boutique of Harrods, the department store mecca, until it closed its pet section in 2014.

Roles for budgies (or parakeets, as they still tend to call them stateside) in American films and television series have been less common but harder to top. Take for example the appearance of actress Anne Francis wearing two live budgerigars in hoop earrings in promotional shots for her role as Altaira in the 1956 science fiction film *Forbidden Planet*.

In the same year Jack Kodell, billed as 'The Original Bird Manipulator' who had followed his father's advice to 'do something different', appeared on *The Ed Sullivan Show* with his famed parakeet/budgerigar act.

And never mind who shot J.R. Ewing, the scheming, greedy villain of the prime-time American soap *Dallas* (1978–91) played by Larry Hagman. Who knew his trademark Stetson, now held by The Smithsonian Institute, featured a band made of dead budgerigars?

Crossing the Pacific, Robert Merlini became the first magician to introduce the early television audience to the budgie in its native homeland in 1957. From that time the budgie flits through the storylines of such classics as *Mother and Son* and *A Country Practice*. Most memorably, in the saucy 1970s soap *Number 96*, when Flo Patterson's bird, Mr Perky, accidentally dies and is replaced, the substitution is discovered after Mr Perky's locum declares 'Hello, Sexy, drop your knickers'. Around Sydney at this time a street performer known as The Birdman could be found busking beside the El Alamein Fountain in Kings Cross with a violin and a troupe of performing budgerigars.

Ironically the budgie was slightly slower to take off in Australia than overseas, with the first dedicated local Budgerigah Club (as it was then still often spelled) not formed until 1932 as part of the avicultural section of the Royal Zoological Society. Up until then budgies had been an adjunct to the canary, cage bird and pigeon clubs which had existed since the mid-1880s. Again, it was the well-heeled at the forefront of the fancy (as the bird-loving fraternity was known) with Warwick (Oswald) Fairfax, then managing director of the family media business, being one of the best-known breeders of the 1930s. His uncle J.H. Fairfax hosted the second lawn show of the now Budgerigar Club of New South Wales at the family's Double Bay pile in 1935.

The budgies' second wind in the United States, where they were reckoned to number 16 million by 1958, was partly due to the number of servicemen who took home the 'lovebirds' at the end of World War II. It was quite astonishing how many birds were transported by servicemen generally. Aboard the aircraft carrier HMS *Centaur* in 1955 were 570 budgerigars, brought to England from the Mediterranean by crew members taking advantage of the end of the UK parrot ban.

In Canada the budgie population jumped from an estimated 5000 to 1 million in the decade after the war. One company that had started out as a small-scale animal food supplement supplier, then moved into birdseed and bird wholesaling, grew from a business with a handful of employees to a huge operation with a staff of 150, with aviaries and branches around the country.

By then there was already a huge divergence between birds so carefully bred for showing and those being sold around the world for the pet trade. Australia now imported 'English budgerigars' that their own great-great-great grandmothers wouldn't have recognised had they alighted heftily alongside them on a branch in the mallee scrub.

Chapter 17

Budgezilla

WEIGHING IN AT 90 grams on the scales, Georgie the budgie was indeed somewhat porgie, but his size owed as much to sunflower seeds as pudding and pie. Obesity is a common malaise among birds kept in gilded cages and fed like Nubian kings on high-fat diets foreign to their metabolisms. And, just like people, these creatures will sometimes compensate for unhappy circumstances—such as the lack of a suitable love interest, or of friends of the same feather—by compulsively overeating.

Before obesity became prevalent among people in western countries the consequence of starchy, fatty diets was evident among their increasingly pampered pets. In the 1960s, Cyril Rogers, one of the recognised experts, warned that there were too many Billy Bunter budgies in Britain: 'Owners are killing their birds with kindness. Do not let them eat cake. It's quite wrong for them.'

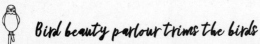

Bird beauty parlour trims the birds

LONDON—Slenderizing courses for budgerigars is the latest innovation offered by beautician W. Farquhar-Moody at his Birds'

Beauty Parlour in Slough, England. Feathered customers come to him for beauty treatment from all over the country. Regular clients have a monthly preen and powder, while their owners may go down the street for a hair shampoo of their own.

Budgies, canaries, parrots, pheasants and strange foreign birds all can be made to look their smartest. And, if William finds a budgie is over the 1½ ounces it should weigh, it's marked for slimming treatment. Its owner will be told to get a larger cage so that the bird can get off the superfluous portion of an ounce. Ten minutes in a flight cage, a quick powder spray, and the bird goes back to its owner.

—*The Tampa Tribune* (Florida), 31 May 1959

Fat budgie syndrome often leads to hepatic lipidosis (fatty liver disease), which is one of the more common nutritional conditions vets see in caged birds. It has led to the development of a pretty radical form of weight-loss surgery for budgies (and other birds) of which Dr Espen Odberg is the proud pioneer. His work has been published in scientific journals and in the original bird treatment bible, *Clinical Avian Medicine and Surgery*.

The Norwegian exotic veterinary specialist performed his first budgie fat reduction operation in his Oslo practice in the late 1980s. 'There are a lot of budgies in Norway. We saw maybe five a day,' Odberg tells us (now back in his homeland on sabbatical after a five-year stint in New Zealand). 'Most of them are suffering from being overweight and the reason is they are just eating seeds and not even a variety of seeds. Because there is so much supplied they can choose the seeds they want.

'I am a pioneer and, I guess, I am the only one that has produced articles about removing fat from budgies. I am quite sure I am the person in the world who has operated on the most fat budgies.'

It is not uncommon for birds presenting to Odberg's practice to be carrying the weight of a whole extra budgie in fat. 'If the normal weight is, say, 40 grams, they could weigh 80 [grams] which makes it impossible to fly, which also then gives them other problems with their feet,' Odberg explains. Because it is almost impossible to put an already obese bird on a diet, drastic surgical intervention is required. As Odberg puts it, 'many of them would prefer to die instead of trying new food. They are refusing the food and after three to four days this becomes critical and will kill it.'

The operation involves removing the subcutaneous fat between the muscle and the skin where they develop large lipomas [growths]. This is not something you would just ask your local vet to do. 'You need to have done many of them to accomplish this but, done properly, it gives immediate results,' Odberg says. 'The bird can fly again and that is good because then it can exercise.

'Sometimes in order to treat a bird you have to do several operations because you can't remove too much fat at once. You may also need to remove some of the skin because the bird is self-mutilating because the skin has too much fat in it and is annoying him so he is eating his own flesh.'

Odberg, who estimates he has done several hundred budgie fat reductions, has carried out a similar procedure on finches, which can weigh as little as 8 grams without surplus fat deposits.

How do you operate on birds so small? 'Very carefully,' says the vet surgeon with a droll chuckle. With a half-hour operation of this kind costing between $400 and $500 it represents no small commitment from the owner.

These are not people who are deliberately negligent, Odberg believes. 'The problem is the people who have exotic pets think they are self-sustained. It is because they don't know better. They are providing food and they think that the animal is able to find its way.'

Imagine: John Lennon's fat budgie

John Lennon wrote a nonsensical poem called 'The Fat Budgie' that was included in his second book, *A Spaniard In the Works*, in 1965. The original handwritten version of this poem sold for US$143,000 at auction in New York in 2014.

Lennon allowed the accompanying 'Fat Budgie' illustration to be used by Oxfam for the charity's 1965 Christmas cards at the height of Beatlemania. The front of the cards contained a facsimile of Lennon's signature and sold in record numbers.

Lennon's widow Yoko Ono granted permission for Oxfam to reprint the card in 2007 to mark the 50th anniversary of the first release of their seasonal cards, which have raised more than $75 million for the charity since 1957.

The famous sketch was part of a collection of works given by Lennon to his editor Tom Maschler. It was sold at auction to a private collector in 2017 for almost US$30,000.

Lonely widows raising little bird Buddhas in one-budgie-policy households from Lillehammer, Norway, to Lepperton, New Zealand are just one end of the human–bird spectrum. At the other end of the scale there are those top-flight birds you won't get out of their nesting boxes for less than $1000.

No species on Earth has had such close attention paid to its genetics by so many as the budgerigar. The modern exhibition budgie strutting its stuff on the show bench represents more than 150 years of selective breeding, including inbreeding or line breeding, to conform to avian supermodel standards.

The exhibition or show-bred bird is more than twice the size of the original wild type and also larger than the pet-shop-variety budgie. Wild budgies average about 18–20 centimetres in length and weigh about 25–30 grams, pet shop birds tip the scales around 40 grams, and exhibition birds weigh 60–70 grams and have an ideal length of 24 centimetres from the crown of the head to the tip of the tail. The competition birds also have larger heads, fluffier feathers and a dazzling array of colours and varieties, with more than 30 primary mutations. Due to inbreeding both deliberate and accidental, other mutations can occur that are quite sad and freakish.

Famous budgies: WHIPPER AND HAROLD

When a budgie is rejected by its parents, it is often because they know something is wrong with their offspring. So it proved with Whipper. Long before it became evident to the human eye that Whipper was a mutant, his mother knew something was up and twice threw him out of the nest after he hatched in 2011.

New Zealand breeder Julie Hayward took pity on the naked, blind baby and became his surrogate, but it soon became clear Whipper was a bird of a very different feather. A profusion of curly plumage grew from his body, caused by what is known as the 'feather duster' mutation.

Birds carrying this genetic defect seldom live beyond the first twelve months and only survive this long if they are given a lot of extra care. The birds are literally killed by the overgrowth of their own feathers. They cannot eat enough to support the feather growth, which robs the rest of their body of nutrition until they die of starvation or exhaustion.

A budgie from Gympie in Queensland, Harold, was the longest-living feather duster budgie ever recorded. His owner, Crystal Gibbons, made the decision to have him euthanised at seventeen months as his quality of life had begun to deteriorate. Harold was a character who used to entertain visitors to the Gibbons's pet shop by playing with his ball. However, his survival was only possible due to the close and loving care he was given by Crystal and members of her family.

'We would cut his feathers one to two times a week. We used to actually have to soak him for at least half an hour a day and then we would trim him because the feathers would just keep growing. We would cut the feathers around his eyes back at least once a week so he could see and didn't get infections.

'But what started happening was his feathers started to grow even more rapidly and that stripped a lot of energy from him and no matter what dietary supplements we gave him, we just couldn't keep up with it.

'We put him down because he got to the point he couldn't play anymore and we knew it was time. It was really sad. He was such a sweetie.'

As the world's most competitive species, with millions of exhibition budgies bred each year, the fancy also lurks in the background of some big personalities. Well-known Australian television vet Dr Harry Cooper and his childhood mate and erstwhile gardening presenter, Don Burke, both took to breeding budgies as kids and continued the interest throughout much of their careers. While Cooper has segued more to poultry in recent years, Burke remains active in budgie circles as a breeder of heritage and miniature budgies. In 2015 Burke, with fellow aficionado Bob Pitt, established the Australian Heritage Budgerigar Association to preserve some of the early colour varieties developed in Australia before 1962 which 'in their original form teeter on the edge of extinction'.

The goal is to develop a 'living budgerigar gene bank' to preserve early varieties in much the same way as other organisations seek to preserve the gene pools of heritage sheep, cattle, pigs and poultry that have been overtaken by more commercial breeds.

The most successful modern exhibition breeder in the budgie's native land is unquestionably Queenslander Henry George OAM, who was the first inductee into the Australian National Budgerigar Council (ANBC) Hall of Fame and has produced more than 43 national competition-winning birds across fourteen different classes since 1982. At the 2015 national titles held in Mandurah, Western Australia, he won an unprecedented seven ANBC 'Logies' with his birds claiming top prize in seven classes—a feat many in the fancy believe will never be repeated. George has also judged the nationals six times and the fourth World Budgerigar Organisation championship show held in Lisbon, Portugal, in 2016.

George was a relative latecomer to the world of budgerigars, which are at the opposite end of the livestock scale to the warmblood horses bred by his wife Diane on their property in the hills outside Brisbane. It began modestly, as these things often do, with a single pair of birds in a cage, and now extends to a purpose-built aviary as big as a house with generous flights for each sex and even a special wing for the aged pensioners who have moved beyond breeding and showing.

'I had two daughters, both of whom wanted budgies, so we went out and got a yellow one and a blue one and they started to pair up, so I went to the pet shop and got a little nest box,' says George of his origins in the field. 'I said to the girls, "Well, we have babies coming—what do you want to do? Do you want to give the babies away to your friends, do you want to sell them for more pocket money, or should we build an aviary in the garden?" Well, they went for an aviary in the garden. That was the start of it.'

When Australia lifted a 40-year embargo on the importation of birds and eggs in 1990, George was among the first to bring in birds from Europe. The introduction of new genetic material made a huge difference to the fancy in the budgerigar's homeland, where the Australian show bird had developed in isolation as a class of its own.

'The Australian budgerigar at that time was a smaller, finer-feathered bird. The imports introduced the bigger birds with longer, thicker feathers, bigger head feathers. Basically, the Australian bird ceased to exist on the show bench and the European bird took over.'

Today George's stud is based on bloodlines from the champion German breeder Jo Mannes, who was for many years considered the best in the world, and lines from other legendary English budgerigar luminaries including Pilkington, Binks, Sadler, Topliss, Wheeler, Wheatley, Norris and Cox.

With hundreds of birds in their aviaries, breeders very seldom name individuals. The exception in George's case was James Bond, so named for his 007 ring number. A blue and white pied of exceptional quality, he won shows and has actually produced three national winners. 'He is seven years old now and probably getting a bit past it. I have a smaller flight for the geriatrics. I should probably be in there myself.'

Another Hall of Famer is South Australia's Kelwyn Kakoschke, who is something of a rare bird himself, perched in his architect-designed house that is literally surrounded by aviaries.

Kakoschke was nine when he was given a little aviary containing six budgerigars of varying colours for Christmas. 'It wasn't very long before I was thinking why are they different colours and that sort of led me into breeding.' By age eleven Kakoschke had his first champion at the Royal Adelaide Show, and a childish hobby soon became a serious pursuit. 'When I started breeding as a kid there wasn't a lot of difference between the backyard bird and the show bird. Now they are worlds apart.'

With more than 1000 trophies for prize birds in boxes under his house, Kakoschke is unimpressed by what passes for standard in many parts of Europe. Quarantine laws effectively mean Australian birds can't compete with the rest of the world, but if they could Kakoschke is confident the home team would give many of the overseas breeders a run for their money. 'My birds conform to world standard where a lot don't conform to any standard,' he says. 'You see these birds overseas that are sort of slumped like old men because they haven't got the bone structure, the line or the vitality. They haven't got a clue what is happening around them because they can't control their feather.'

Kakoschke likens budgerigar breeding to genetic sculpting. 'I guess what you are doing is mixing science with art and that is really my background. You are not just looking for flash-in-the-pan mutations. You are sculpting through inbreeding, line breeding and careful selection to get a bit more back skull, a bit more directional feather or to get that back line a bit more perfect.'

 ## Strike me pink!

The pink budgie has long been the unicorn of the bird world and the stuff of marketers' dreams. No one has yet produced successive generations of a conclusively pink bird, though many have tried. Throughout the 1950s, when the fancy was at its pinnacle, there were reports of red, pink and even black budgerigars bobbing up in aviaries in Australia and overseas.

In the early 1950s a report of a pink budgie on display in a city in the UK brought fanciers from across the country. But the famous bird, bred by an amateur, soon began to fade and it transpired the owner had been feeding the creature pink dye used in the marzipan factory where he worked.

Over the years pink calcium perches, Condy's crystals, iodine blocks, cochineal, prawns and even beetroot have all been found to play a part in colouring pink or red birds, leaving their owners, who in some cases genuinely believed they had produced the holy grail, blushing.

The one case that was authenticated—according to John Scoble, the author of *The Complete Book of Budgerigars*—involved a Melbourne breeder named Ron Jones. He was reported to have produced twenty of the birds with the depth of colour improving. Unfortunately, a newspaper report about the birds also revealed where the breeder lived and they were all stolen from Jones's aviary, never to be heard of again.

Professor Harry Harrison, who wrote himself into aviculture annals by designing a computer program to optimise exhibition budgerigar breeding, was thought to stand the best chance of claiming the £1000 offered by a birdseed company in the 1980s. The professor, who was an avid budgie lover and a keen exhibitor himself, hoped extraordinary binary powers would eliminate much of the trial and error, and would pair off the parents of the first pink budgerigar. It never happened!

Betting company William Hill for a long time have offered odds of 100/1 for a 'small fun bet' for those who fancy a flutter on the likelihood of a pink budgie. The bookmakers have taken a few bets in years gone by and there have been a number of fraudulent claims.

'Producing an indisputably pink budgie is marginally less likely than the Loch Ness Monster turning up, but technological advances used with humans could be used to breed a pink budgie. That's why the odds are 100/1 and not 1,000,000/1,' a William Hill spokesman commented almost twenty years ago. 'We haven't laid a bet on this for over a decade, mainly because no one has asked us,' spokesman Rupert Adam says today.

The most frequent misrepresentations and false sightings have been caused by a bird that is not a budgie at all but a Bourke's parrot, according to the Budgerigar Information Bureau, which offered £500 for the first genuine pinky back in the 1970s.

However, there is one Burke in the pink budgie camp who is no imitator. Don Burke has made no secret of his campaign to breed the world's first pink budgie. Writing in *Budgie News Victoria* in 2018, the former television presenter outlined his plan to breed pink budgies using some 'reddish-purple heritage violet budgies'.

William Hill may have a winner on their hands.

Chapter 18

It's showtime!

THE YEAR IS 1956. Elvis Presley's 'Heartbreak Hotel' is top of the hit parade and the teenagers are scandalising polite society with 'rock and roll riots'. Melbourne is about to host the first Olympics ever held in the southern hemisphere and this very grown-up moment in Australia's history will be captured live on television, but Charlie Bush has other things on his mind.

The burly garbo from Fitzroy North is chuffed the forecast rain has held off as he pushes a pram full of budgies and finches up the hill along St Georges Road. The missus and a gaggle of tow-headed kids lend an air of procession to the 2-mile walk to the Dispensary Hall in Normanby Avenue, Northcote.

Others have cycled to the Northcote Cage Bird and Budgerigar Society show with cages of birds strapped to their backs, bike racks and handles. Cars have not yet forced street cricketers off the roads in working-class suburbs, although Clydesdales have already gone the way of the night cart man. In just over 100 years the world's most democratic parrot has passed from the hands of bluebloods to everyman.

Alan Rowe, a plumber's apprentice, was already at the hall when the Bushes tumbled noisily in and the serious bird-judging business began. It would be six years before young Rowe took home the coveted Parson's Trophy for the best hen in show with a lutino, an all-yellow budgie characterised by the absence of dark markings and pink/red eyes caused by the Ino mutation.

Rowe first got hooked as a sixteen-year-old, seeing 'all these colourful little budgies with their big ribbons' at the Royal Melbourne Show. By 24 he had become a club show judge himself, helping select which birds would win the coveted rosettes for most closely adhering to a strict set of standards for size, shape, colour and condition.

Now nudging 80, Rowe has the equivalent of a black belt or grandmaster title in budgies, having clocked up sixteen national championship class wins, and then going on to judge a number of national shows—six of them in Australia, as well as others in New Zealand, Germany and England as part of the World Budgerigar Organisation Panel of Judges.

In 1994 his achievements earned him a place in the ANBC Hall of Fame and, although he has now retired from judging, he remains very competitive as a breeder. The appeal, Rowe informs us, lies in the challenge of trying to breed birds that most closely meet the standard. 'I probably don't get as much pleasure out of how they act and carry on now like I did when I first started,' he admits. 'I look at them with the standard in mind.'

Presently he is preparing birds for selection for the United Budgerigar Society's championship or diploma show, with the top three birds in each class going on to vie for a place on the team to represent Victoria at the Nationals.

'It is quite a process,' Rowe explains. 'You have to think about getting them into condition well before the show. It takes eight weeks for them to grow a tail so if they have a broken feather or that sort of thing you remove those.

'In the last few days I have sprayed them to get rid of any lice and mites and also treated them for a disease called canker as a preventative measure.' Closer to the show he will spray them again and pluck out any extra black spots on their face beyond the requisite four main ones and those under each cheek patch.

The morning of the show Rowe goes to the bird room early and turns the light on to wake the chosen contestants so they can have a feed and don't spend the day hungry. 'I don't wash and blow-dry them, although some people do,' he says. 'It makes their feather boof up more, but I am a bit scared about cooking 'em so I don't do it. Plus, I don't imagine they like it very much and there's enough stress with the transport and everything.'

Rowe has a couple of promising babies from his 2018 winner in the recessive pied class and has his fingers crossed they will get across the line. But show business, even among budgies, is highly competitive.

His club of about 30 members, including fellow Hall of Famer Jeffrey Leong, who clocked up the requisite seven national wins in four shows after switching from pigeons to budgies in 2009, currently holds all three state championship shields. As a particularly strong club, based in Melbourne's north-eastern suburbs, United is usually well represented in the Victorian team, which has won 24 of the last 30 Australian National Budgerigar Championships.

All sixteen clubs in the state are entitled to three birds in each of the 28 colour and variety classes, which equates to more than

1300 birds and a tremendous volume of chirping on the day as they perch in their identical show cases. The contestants in each group of 45 are judged down to twentieth place and the points then awarded to the club. The club that wins the most points claims the shield and the championship. The three big shows are for 'unbroken caps' or UBCs, as juvenile birds before first moult are known, young birds aged six to eighteen months, and adult birds.

Like horses, all budgies have the same birthday, moving from young to adult category on 31 December of the year after they have their legs ringed by breeders at around ten days old. The numbers and letters on the rings identify the bird's sex and variety, and the breeder's club affiliation.

It is the goal of every budgerigar breeder to produce a 'stormer'— an excellent show-stealing bird that nails the standard from the get-go and presents in tip-top condition at just the right time. Judges carefully assess the size, shape, back line and condition of the bird as well as the way they show. If they don't relax in the show cage, then puff their feathers up and confidently come to the front of the cage, they won't cut it.

The 'modern-look' bird is a far cry from the days when Rowe started out in the fancy. 'It was really a time when you had to make your own fun,' he recalls of his teenage years. 'In those days people also used to trap finches and sell them so we did a little bit of that to get a bit of pocket money. Another of the things you could do is have pigeons or budgies. Canaries were out because they were a bit dearer and a little bit more finicky as far as getting them to breed. The budgie clubs were also a bit of a social outlet and were pretty popular.'

Back then dealers were still getting in boxes of wild budgies. 'Every now and then there would be a couple in the box that were a little bit bigger and showing bigger throat spots and were just a bit different and you'd grab those'.

There were a few different bird dealers in Melbourne, Rowe says, but a dealer named Alan Beattie stood out. 'When people culled their birds they would take them to him and if you had an eye for them and got there at the right time you could pick up good birds without paying big money'.

 ### Bird fancier's triumph

Breeds blue budgerigar with yellow wings

MELBOURNE, Saturday—E. Anderson, a 21-year-old factory hand, of McDougal-street, Geelong West, is believed to have achieved the objective of thousands of bird breeders by producing a blue budgerigar with yellow wings.

Fanciers have attempted to obtain these results for years, but Anderson has definitely established the type with three male birds raised in his aviary. He intends to perfect the colours before putting the birds on the market.

Should Anderson's efforts be successful, he will be able to command his own price for the birds.

When the cinnamon-winged budgerigar was produced its owner made more than £700 in one year. . . .

'It was largely luck getting the color,' said Anderson to-day, 'for the parents were both yellow birds, but the different breeding, with blue and cobalts in earlier matings, produced the result.'

—*The Sun* (Sydney), 19 January 1936

After the young plumber got married and started his own family, the budgies soon took a back seat to his own children's interest in the pony club. 'Once I started working for myself, I got fairly busy and the budgies were only an outlet from work. I wasn't really professional in my outlook at the time.'

Later, when his children outgrew the pony club, Rowe decided it was time to get serious about the budgies. When the door once again opened to import parrots through the Spotswood Quarantine Centre, he joined one of the syndicates formed to bring birds in from Europe.

'We had an English breeder come out to our club and he acted as a bit of a go-between for us,' Rowe explains. 'We went over to select the birds in 1989 and he took us to various places. Some were a lot dearer than the others. The dearest one I got was £200 at that time. Each bird ending up costing around $500 on average by the time you paid their quarantine. A lot of people only brought in half a dozen or four birds. By the time they got here some of them were past their prime and a lot of them didn't breed.

'Some of them put them straight into breeding boxes as soon as they were out of quarantine. I flew mine for a while before I bred with them and worked in with another guy who also brought in about fifteen birds. We worked in together for two years to get them going. That gave us a variation of breeders and gave us a good start because we bred them up in numbers.'

The influx of European birds had an immediate and dramatic impact on the competition. 'Suddenly we had birds exactly the same as over in England,' Rowe explains. 'I had only won one National before then, and the rest of the Nationals have been won with those birds' bloodlines. The birds have improved every year since then. The birds got spread around and the people who couldn't afford to import finally got their hands on them and we have some very good budgerigar breeders in Australia.'

Since Australia became a member of the World Budgerigar Organisation in the late 1990s it has adhered to the world standard of perfection despite not actually being able to compete against other nations because of quarantine restrictions. The key difference between the Australian national competition and the shows in Europe is that there is no one grand champion here. Breeders strive for excellence across all classes.

As Colin Flanagan, president of the Budgerigar Council of Victoria, explains, 'The English people are trying to breed the best bird in the world. In Australia we are trying to breed the best bird of a variety, be it clearwing [variety with no markings on wing], lacewing [very light markings on wings], spangle [fewer markings on head and shoulders and body colour extends into the wings] or black eyed. Those varieties of birds and the definition within those varieties are the best in the world and that is where we have separated from the UK.'

An accredited judge and early Hall of Fame inductee, along with his breeding partner Bruce Sheppard, Flanagan is another who has kept budgerigars since his early teens. Bringing his CEO skills into play, the business manager believes the organisation needs to better market itself and improve its service offering so it represents the entire fancy, from top-line breeders and keen exhibitors to social members, and those who concentrate on certain types such as wild budgies or rainbows. 'Obviously they can't be judged against the stuff we have got because they are chalk and cheese, but I believe we do need to be more encompassing of the entire fancy, to add some classes and draw people in more.'

The individual clubs that make up the Australian National Budgerigar Council have become more professional and standardised in their approach, much to the relief of successful budgie breeders' spouses whose homes were at risk of being overtaken by dodgy objets d'art awarded as trophies.

In the office next to the bright, clean budgie room where Rowe can house up to 60 breeding pairs, there is row after row of trophies, from the standard 'Logie' to all manner of vases, paintings, goblets, bowls, clocks, trinkets, plaques and boxes of ribbons awarded as prizes. 'When the Australian titles started the clubs used to have their own trophies and it became a bit of a deal because some of the clubs in richer states used to have better trophies than others,' Rowe recalls. 'I have a few of those early ones that are my best possessions with budgies because they were quite nice, but some of them were truly shockers.'

 ## Show budgerigars 101

There is a dizzying array of budgerigar varieties. 'There are colours and shades including light, dark and olive green moving along the spectrum to grey green, through light, dark, cobalt blue and grey into violet,' explains Vic Murray, education officer of the United Budgerigar Club. 'There are also colour varieties— English yellowface and Australian yellowface. Then there are the genetic varieties available in all the colours. Some of the colours and varieties are genetically dominant, some recessive, and some sex-lined and carried by the cock bird.'

With 32 primary mutations enabling hundreds of second-ary mutations, both stable and unstable, there's a lot to learn for beginners. The most recent mutation, which in 2019 made it to the program of the Australian Budgerigar Championships as a demonstration class, is the white cap which, as the name suggests, is a bird with a white forehead.

The mutation was first identified as such in a Queensland aviary in 2003 and caused considerable excitement among breeders, being the first Australian variation to emerge since the 1970s. The white cap remains an exclusively Australian mutation at this time.

There are other mutations that have occurred in Australia and been recognised by the World Budgerigar Organisation, thus becoming part of the international show scene.

The crested budgerigar is one example. Budgerigars with crests were first recorded in Australian aviaries in the late 1920s and were among stock exported to England. From there they spread to the continent, Canada and the US.

The Japanese have been at the forefront of the development of newer crested varieties, producing birds with additional crests on their wings, like birdy shoulder pads, in the 1980s. These birds, known as Hagoromo or helicopter budgies, tend to be less popular in the West, but are highly prized in Asia and the Middle East.

Chapter 19

Smuggled budgies and dark deeds

THAT LATE GREAT comic genius John Clarke is credited with coining the profoundly ego-withering term 'budgie smugglers' as a descriptor for the type of swimming briefs favoured by surf lifesavers and erstwhile Australian prime minister Tony Abbott. The phrase was first publicly uttered in series 1 of the ABC television mockumentary *The Games* in 1998 by the character played by Clarke, and it seemed to become entrenched in the national vernacular almost overnight.

As he offered with that trademark droll economy: 'Des Renford [legendary distance swimmer] would regularly take on the English Channel, Bryan. He would drop his tweeds, pull on a pair of over-sized budgie smugglers and he would drop a bomb off the white cliffs of Dover and start rolling his arm over.'

In 2016 this glorious phrase found its way into that great arbiter of the lexicon, *The Oxford English Dictionary*. (Later this same year nine young men—the 'Budgie Nine'—stripped down to budgie smugglers sporting the Malaysian national flag at the Malaysian Grand Prix in a stunningly ill-conceived show of support for fellow countryman and winning driver Daniel Ricciardo. The stunt landed them briefly in gaol.) Oxford Australia had previously identified the highly descriptive compound noun as the word of the month for December 2010, hypothesising the term may have been influenced by 'grape smugglers', an earlier term for men's swimming briefs used in 'international English'.

It's just as plausible, though, that Clarke recalled a case involving the real thing that occurred back in his native New Zealand. Clarke was still farnarkling around at university in Wellington in 1968 and yet to really work up his Fred Dagg character after a stint as a shearer, when customs officers smashed an organised racket smuggling parakeets, finches and budgerigars across The Ditch.

To highlight the devious methods being used by smugglers, the NZ customs minister at the time, Norman Shelton, famously brandished a giant pair of men's green underpants in which birds had been concealed in specially constructed pockets.

The smuggling of native, wild and exotic birds, particularly parrots, is a long-standing worldwide problem. In 2010 almost 248,000 psittacines (parrots) taken from the wild were imported by the European Union legally, but many birds cross world borders illegally beyond the huge captive-bred trade. The worldwide wildlife black market, in which parrots feature heavily, is reckoned to be the third-largest illegal trade after drug and gun trafficking.

As soon as Australian states began introducing ordinances in the late nineteenth and early twentieth centuries to curb the depletion of native birds, smugglers sought to find ways around them. Newspaper articles dating from the 1920s and earlier lament that 'acts of parliament are no proof against the unscrupulous trader'.

'Japan and China provide a ready market for Queensland's painted finches and budgerigahs,' *Smith's Weekly* reported in 1927:

Absolute protection or closed season, it is all one to the bird thief, however. When he can get up to £6 and £7 a pair for finches or budgerigahs [sic] he does not bother about regulations . . .

When boats from the Orient are in port a search is made by the authorities as near to sailing time as possible. Birds have been found concealed under hats, in coat pockets and in small cardboard boxes attached to belts.

Nor does the ingenuity end there . . . a vigilant Customs officer noticed that certain members of the crew of an outward bound Japanese steamer were wearing boots which appeared to be several sizes too big for them. A search was made and it was found in the toe of each boot a finch, wrapped around with wadding had been concealed.

After the Parrot Fever outbreak of the 1930s port security was ramped up a notch, but there were still people willing to chance it, like the liner passenger caught when his shirt 'twittered' as he attempted to smuggle six budgerigars into New Zealand in 1937. There were also those occasions when people attempted to smuggle exotic parrots from the Americas and Asia into Australia.

 Customs man hears cries for help

Tell-tale parrot outwits bird smugglers

Muffled cries for help which echoed through the darkness at Victoria Docks while the liner *Nestor* was in port led an alert Customs officer to the gangway, where he intercepted two firemen as they stepped on to the wharf.

'Help! Help!' The cries seemed to come from inside one of the firemen's coats.

Budgerigar

It was a green African parrot which was making an appeal from its binding of an old flannel shirt. A bulge in the other man's pocket was made by a beautiful Amazon parrot.

The two birds, which were valued at more than £30, were confiscated and sent to the Melbourne Zoo, where they are now being kept. A search of the ship revealed other undeclared birds, which were also taken for the Zoo.

Recently a number of Eastern birds was taken by the Customs officials from the *Nankin*. Because of parrot disease precautions, parrots are not permitted to land in Australia.

—*Evening News* (Rockhampton), 11 March 1933

But in the case of the budgerigar, unlike many other parrot species, it wasn't the original wild-caught bird that people wanted. Even before the turn of the nineteenth century the preferred target of thieves and smugglers supplying the private collections, private zoos, pet trade and breeding farms in America and Europe was birds with colour mutations produced by decades of captive breeding.

Due to an acute shortage of locally bred 'parakeets' in the US in the 1950s, even common green and blue varieties were being onsold to retailers and brokers for as much as US$15 each, while rarer coloured varieties could fetch up to US$200 per bird. With that kind of financial incentive, budgies were crated up and flown into Mexico before being smuggled via mountain trails across the border into the US. Customs officers of the day described how the budgies were starved for 24 hours and then fed on whisky-soaked bread to keep them sedated during the overnight trek on the back of wildlife mules. Once over the border the birds were loaded onto trucks and taken to Los Angeles bird farms from whence they were sold and distributed through normal retail outlets.

In Australia, where members of the fancy were cut off from the array of genetic mutations being cultivated through decades of selective breeding across Europe, show birds really were in a class of their own. But somehow it never seemed to take long for some of these more desirable qualities to appear, fully fledged as it were, in local aviaries. It was no secret that European birds with coveted qualities were being smuggled into Australia and this became one of the main arguments for the lifting of the 1956 embargo to allow controlled importation. The ban ended in 1990.

Of course, smuggling is not the only form of skulduggery known to occur in the bird world. Budgies can attract big bucks, with a record £10,000 paid for a single bird in the UK. Even in the much smaller, less hotly contested Australian market the little creatures sometimes exchange hands for extraordinary sums, with a record price of $8500 paid for a single bird in 1992 and another fetching $6000 at a sale in Penrith as recently as April 2019.

As with the competitive dog and cat show scenes, there is a muddy underbelly to the international budgie business, with dark tales of feather snipping, plucking, painting, dyeing, gluing, steroid feeding, theft, tampering and even murder. Accounts dating back as early as 1915 reveal how unscrupulous breeders would burn or apply brown boot polish to the ceres of male birds so they might pass as more valuable hens.

Gerald Binks, the late, legendary British fancier, judge, author and founder of *Budgerigar World* magazine, used to tell of one competition back in the 1960s when the tails of several of the best birds were snipped off the night before judging so they were not complete and were disqualified. Strong suspicion fell on the competitor who walked away with the first prize rosette on that occasion. He was a hairdresser and his daughter, as organiser of the show, had a key to the hall.

These days security is taken a lot more seriously, with compulsory leg rings for show birds, plus a clear chain of custody and iron-clad protocols for competitions.

 ## Police chief's 'budgie' stolen

Among the list of unsolved crimes in the Brierley Hill police records there is now an entry which is worrying all the force—the theft of the chief inspector's budgerigar.

Chief Insp L.H. Lockett's house in Delph Road, Brierley Hill was left unattended for an hour during the mid-week and when his son returned he found that a pane of glass had been broken in a window, the door was open and the budgerigar, with its cage, was missing.

Nothing else had been taken.

—*Birmingham Daily Post*, 14 November 1956

But the increasing sophistication of aviary theft remains a recurrent, cyclical problem in the UK, where it tends to rate rather low on the list of police priorities—as Andy Pooley discovered. A former quarry worker with a wrestler's build and a velvety West Country accent, Pooley was utterly devastated by the theft of 21 of his top birds when he left them briefly unattended the night before the Cornwall Budgerigar Show in August 2010.

What hurt him most was that his very best bird, which had won grand champion at the same show a year earlier, was trampled to death, apparently in the thieves' rush to remove the birds from the aviary. The bird, named Penmead Pride, was Pooley's pride and joy and, as a registered champion, more valuable than the combined worth of the stolen birds. Because of the time frame and 'murder' of Penmead Pride it was initially thought the theft may have been instigated by a UK rival, but after Pooley enlisted the aid of a private detective it seemed more likely that the birds were destined for the European market. No trace of the birds, at home or abroad, has ever been found.

'It was the worst thing that ever happened to me in my life. It just broke my heart,' Pooley says. 'I had my birds stolen on a Friday evening and I phoned the police and they didn't arrive for two weeks. People in the budgie world said they were probably taken out of the country within hours rather than days.'

In his grief Pooley, who was secretary of the Cornwall Caged Bird Society and had been breeding and showing birds since he was in his teens, left the fancy and got rid of his remaining birds.

Now 67, he has only just recently bought himself a couple of pet budgerigars. 'I had a second heart attack and got fitted with a defibrillator and I was getting very down,' Pooley says. 'My wife said to me, "You need to get some birds again." I used to wake up in the morning and hear the birds singing out the back and I missed it something terrible. So I have got a few now, but I will never show again. I used to love it, but it is too much for me. I don't want to be involved in it anymore. It is all about the money.'

Three years later there was another big budgerigar raid on a home in Hampshire where thieves stole a staggering 350 show birds worth an estimated £60,000. The birds' owner, Mick Freeborn, a national judge and internationally regarded breeder, had carefully developed his bloodlines over 50 years since he'd acquired his first pair. He described the birds as irreplaceable. 'It would have taken three or four people at least three hours to remove them and a van to take them. It is possible they were stolen to order.'

In Australia aviary theft is less of a problem, but that is not to say it does not happen. As the best breeder in the country, Henry George counts himself fortunate not to have been hit, but one of his best mates in the fancy who hails from Townsville has suffered thefts on two occasions. Similarly, people have every expectation that the most recent Australian mutation, the white cap, will soon be appearing in the aviaries of Europe, most likely thanks to smuggled genetic material if not birds themselves.

Occasionally budgies have been an effective means to an end for home invaders and violent thugs who have exploited their victim's love of a pet to extort money. In 2003 Mrs Ethel Price, 86, handed over £500 she had saved to buy a headstone for her husband's grave after two men posing as gardeners entered her Hampshire home and threatened to kill her budgie with a hedge trimmer. 'One of them kept revving the trimmer behind me. I was petrified. They kept looking at my budgie Peter and saying how terrible it would be if something happened to him. He's been my only companion since Dave died,' she told the local newspaper. The heartening response from the local community was to raise four times the money the ruthless robbers took, which not only paid for a fitting memorial to the late Mr Price but also a major garden makeover by a genuine landscaper.

Other crimes and misdemeanours have involved not the budgie so much as one of its favourite sources of food. In 1959 two men from Cardiff in Wales were each sentenced to nine months' jail after they turned a vacant council allotment into a profitable hemp plantation. The court heard the pair had scattered 6 shillings' worth of budgerigar seed on the disused block, which resulted in a flourishing Indian hemp crop of 36 plants. Owing to a particularly dry, warm summer, their plants 'produced fruitful tops and resinous glands usually unknown outside the tropics'. The crop was so good it was found to be worth £2300.

In a 1971 case a small cottage in the postcard-pretty village of Wincle, Cheshire, was found to be the 'centre of an international drug ring'. After growing cannabis from budgie seed in the garden among the lupins and hollyhocks, the owners had been sending it off to individuals in America, Australia, Lebanon and Greece.

 Bird seed shock

Two elderly Runcorn men became suspicious about strange plants in their gardens. They cut the plants and took them to the police station after an expert told them, 'these are cannabis plants.'

But Mr Albert Newby, of Stenhills Crescent, and Mr Frederick Snelson, of Hinton Road, are not in trouble.

A spokesman for the police said: 'It appears the men keep budgerigars and they threw some bird seed out of the cages into their gardens. Of course, we won't be prosecuting.'

—*Liverpool Echo*, 12 September 1973

Should you be tempted, it is better to know now that these days birdseed contains sterilised seed processed from industrial hemp which, even if you managed to grow it, has a tetrahydrocannabinol (THC) level of less than 1 per cent, compared to the 4–20 per cent contained in the plant that is smoked or otherwise imbibed. In fact, the hemp in birdseed actually comes from a different variety of the species *Cannabis sativa* altogether.

It would literally not get a budgie high, but at the peak of the US 'war on drugs' in the late 1990s, members of the US Customs Service were busily impounding birdseed with zeal.

Chapter 20

Doing bird and other therapeutic uses

IT CAN BE stated with some certainty that Florence Nightingale did not have a budgerigar. The founder of modern nursing did, however, have a little owl named Athena, which she carried with her in her apron pocket in her lesser-known role as superintendent of the Institute of Sick Gentlewomen.

This was how the pioneer nurse first came to understand the therapeutic value of animals, and birds particularly. As Nightingale observed in her no-nonsense style in her 1859 book *Notes on Nursing: What it is, and what it is not*, 'A small pet animal is often an excellent companion for the sick'—and certainly better, in her view, for the chronically ill patient than the 'chattering hopes and advices' of their friends. 'A pet bird in a cage is sometimes the only pleasure of an invalid confined for years to the same room,' said the legendary Lady with the Lamp who, upon her return from the Crimea, had her own experience of protracted illness and apparently no shortage of irritatingly cheery human visitors.

You probably wouldn't have to be Florence Nightingale to deduce that a budgie was better for patient morale than, say, a pot plant, but so it was proved in a landmark study published in 1975. In a paper titled 'Some recent work on the psychotherapeutic value of caged birds with older people', renowned British animal behaviourist Dr Roger Mugford and his colleague James M'Comisky provided the first real empirical evidence of the value of what has become known as pet therapy. The study of 30 elderly people who lived alone found those who received budgies were generally more cheerful, more outgoing and less inclined to grumble than those who received pot plants, specifically begonias, and than the control subjects who received neither.

The budgerigars became such a great topic of interest among those fortunate enough to receive them that 'they could even displace the monotonous awareness and discussions of past and pending medical ailments'. Reassessed eighteen months later, the subjects who had been given birds were found to be much more socially engaged and have a lower death rate than the begonia beneficiaries.

A follow-up study by American academics, published in 1980 as 'A wine bottle, plant, and puppy: Catalysts for social behavior', found a caged puppy placed on a nursing home day-room table in front of elderly male veterans produced a more dramatic increase in social behaviours than either a wine bottle or an unspecified variety of flowering plant. But perhaps this result might have been slightly different if the wine bottle had been opened.

 ## Bird seed breasts

Sir,— As one who has tried many types of artificial breast in the last three years, may I urge 'Interested Victim' (*Journal*, April 14 p. 862) to give one made from bird-seed a trial, which apparently she has not done?

I followed the instructions given by Mr D. K. Lennox . . . and at long last have found a substitute for the original which has adequate weight, natural shape, and, I may guess even authentic consistency. Furthermore it neither rattles, leaks, nor sprouts, and the family budgerigar has shown no unseemly interest in me to date.

I may thank Mr. Lennox that my children can no longer say: 'Mummy, your chesty has slipped,' and I am most grateful to him for his letter.—I am, etc.,

'BIRD-SEED'

—*British Medical Journal*, 9 June 1956

Since the 1980s there has been a raft of studies to confirm the therapeutic and educational value of dogs, cats, birds and fish in all types of settings from schools, rehabilitation wards and nursing homes to psychiatric institutions and gaols.

Birds and prisoners have long been associated through both myth and actuality. What creature but a bird can bring a word of hope to a prince or princess held captive high in a tower? In the case of Anne Boleyn—the unfortunate second wife of King Henry VIII—the ravens that guarded the Tower of London where she was imprisoned were said to fall strangely silent and immoveable on the battlements as the queen was led to her execution.

The most famous example of avian and inmate affiliations is Robert Stroud. A convicted murderer, Stroud raised 300 canaries and wrote two seminal books about avian pathology in his cell in

Leavenworth Penitentiary in Kansas in the 1920s and 1930s, before he was moved to an island gaol off San Francisco where he became known as The Birdman of Alcatraz. (He was not actually allowed to take his canaries with him when he transferred to Alcatraz, but it sounded better as the title of the celebrated book and film.)

In gaols, just as in homes, the budgie would soon become the top bird, being generally hardier and cheaper than canaries with the added advantage of being able to communicate with co-caged humans. It's not surprising then that there are more budgies serving gaol time than any other species—apart from the humans who are 'doing bird' alongside them. Feathered inmates have even contributed to the colourful glossary of prison slang; 'budgie syndrome' is a term applied to prison bodybuilders who preen and puff at their own reflections.

Allowing inmates to keep budgies can be more than a reward for good behaviour. It can help build empathy, understanding, responsibility, the opportunity to nurture and a sense of purpose even among the hardest cases. Studies at Oakwood Forensic Center, a maximum-security psychiatric criminal facility in Ohio, for example, found inmates on wards with pet birds were less violent and required less medication.

At the end of World War II several rehabilitation hospitals in Alberta, Canada, introduced 'lovebirds' to bring 'interest and purpose' to the lives of mentally and physically disabled veterans. The idea was proposed by a World War I veteran and bird fancier who had spent a good deal of time in hospital himself after being repatriated. Nursing staff reported the effect of the avian arrivals was marked: 'It is amazing how the morale of the whole ward is improved just by having one of these little birds in it.'

In 1959, the imposing red-brick nineteenth-century Sandhurst Gaol, recast as Bendigo Training Prison, became the first Australian gaol to install a budgerigar aviary. A bird-loving local dentist was drafted to run spare-time evening ornithology classes for prisoners after the aviary was stocked with donated birds. 'The aviary brightens the institution greatly,' observed governor Ian Grindlay (who just by-the-by would follow one of his former Bendigo charges, a model prisoner named Ronald Ryan, to Pentridge and was governor there when Ryan, the last man to be legally executed in Australia, was hanged).

The South African gaol where Nelson Mandela spent the last six years of his imprisonment is home to one of the world's most successful and long-running rehabilitation projects using birds. The Correctional Bird Project at Cape Town's Pollsmoor Maximum Security Prison is helping reform even the most hardened inmates as they hand-rear psittacines of all persuasions, from rare Cape parrots, African grey parrots and eye-catching eclectus parrots to common bush-variety budgerigars.

Introduced in 1997, the project—which has led to obvious comparisons to the film *Birdman of Alcatraz*—was in fact inspired by the prison's parole board chairman who was a member of a bird club. Wikus Gresse has long believed animals have the power to help unlock prisoners' potential to reform their own lives and learn skills that can help them on the outside. 'The same hands which committed assault or stole are being used positively with patience and care,' he says.

Prisoners who qualify to join the program are given parrot chicks to hand-rear for ten weeks whereupon they are sold to a waiting list of customers. The money covers the program's running costs and any residual funds are shared between participating prisoners.

In the late 1980s and early 1990s several UK prisons also successfully involved inmates in breeding budgerigars for charities, which distributed them to elderly people for companionship.

Before a review sparked by media backlash over the leniency of treatment of serious offenders, budgerigars had long been on a list of 135 privileges available to Category-A (high-security) inmates, including lifers, in UK prisons. Under the Incentives and Earned Privileges Scheme prisoners who played by the rules were permitted to buy their own bird, but also required to properly tend to its upkeep. One notorious inmate known to keep a budgerigar is the serial killer Rose West who, for a time, shared her Durham gaol cell with Oliver.

Robert Maudsley, a murderer forever stuck with the nickname 'Hannibal the Cannibal' after it was widely reported—quite falsely as it happens—that he ate part of the brain of one of his four victims, has repeatedly begged to be allowed a budgerigar. Maudsley is Britain's longest-serving prisoner, held in a specially built cell reinforced with steel and bulletproof glass that could have been the blueprint for Hannibal Lecter's cage in *The Silence of the Lambs*. He wrote to *The Times* in 2000 asking that the terms of his solitary confinement be relaxed to include a budgie, a TV set and classical music, or to be allowed to commit suicide by cyanide pill. 'Why can't I have a budgie instead of the flies and cockroaches and spiders I have. I promise to love it and not eat it,' he wrote. His request was denied.

Beyond prison walls, budgerigars have proven valuable aids in education. A school in Coffs Harbour, New South Wales, enlisted a teacher's aide budgie named Captain Hook to the delight of special-needs children as early as 1958.

Dr John De Nobile, director of Macquarie University's Primary Teacher Education Program, is an enthusiastic promoter of the study of budgerigars to promote kids' deeper understanding of the living world around them. 'The use of budgerigars as a teaching and learning resource has great potential to engage primary students and provide a stimulus like no other to develop the scientific knowledge and skills required to work scientifically,' he wrote in his 2013 paper 'Working scientifically with budgerigars in the primary classroom'.

De Nobile's research drew on his own experience as a young teacher. 'I had budgies at home and I brought in these two budgies, one green, one blue. We were doing a Living Things unit in Year 3 and it was really just meant to be a decoration in the classroom, but it became a real learning tool,' he recalls. 'The students were so engaged. It was like nothing else I had in the classroom and it was a fluke. I regret I didn't in the following year actually have it formalised as part of the curriculum because that would have given me a lot more evidence of how engaged students get.'

 Novel experiment

Budgerigars to soothe typists

The managing director of the Manchester Works of the British Dyestuffs Corporation is conducting an experiment which must be unique in the history of stenography.

Convinced that the noise and clatter of the typists' room in the works, which accommodates more than 50 machines, must be detrimental to office efficiency and to the comfort of typists, he suggested to visiting experts from the National Institute of

Industrial Psychology that the introduction of a number of budgerigars into the room would improve conditions. It was thought that
the birds' notes might lessen the fatigue caused by the monotonous noise of the machines.

The psychologists agreed that the experiment was worth trying,
and a number of birds have been installed in the room.

In addition, the typists' desks have been painted in various individual shades so that the monotony of office work may be relieved.
—*Sunderland Daily Echo and Shipping Gazette*, 21 January 1937

PS: It was later reported this idea was abandoned after it was
pointed out the budgerigars would likely just mimic the sound of
typewriters and compound the problem.

Budgies themselves have a great deal to offer science. As creatures that are easily bred, rapidly maturing, remarkably obliging
and capable of ongoing learning, they make excellent laboratory
subjects and have been used in many studies relating to hearing,
vocal communication and learning as well as flight navigation.

With its ability to fly at great speed through complex environ-
ments, the ubiquitous Aussie bush bird is helping provide valuable
guidance information for autonomous aircraft, which are coming
as surely as driverless cars. One study funded by Boeing Defence
Australia and the Australian Research Council and conducted by
researchers from the Queensland Brain Institute sent opposing
pairs of budgies flying towards each other through a tunnel to
establish how they avoid mid-air collision. After 102 incident-free
flights researchers concluded the reason was that the birds always
veered right.

Other studies have examined how birds use visual clues to navigate tight spots and automatically adjust their wingspan. One quite incredible test conducted by researchers at Duke University in North Carolina in the late 1990s involved training a budgerigar to fly down a wind tunnel wearing a teensy-weensy oxygen mask to measure how much muscle power it required. As it turns out the budgie punches off the scale in power-to-weight ratio.

While providing information that can help with the design of navigation and guidance systems, and further enhance understanding of the physics and evolution of flight, budgerigars may yield even more important clues when it comes to cognition and ageing, some researchers believe. Canadian biologist Dr Andrew Iwaniuk, research chair in comparative neuroanatomy at the University of Lethbridge in Alberta, has a great deal of time and respect for budgies.

'Birds have long been a model in neuroscience for studying learning and memory,' he says. 'Parrots, including budgies, have this ability to learn new tasks and do different things and essentially maintain flexibility in their behaviour that is far more like humans than using a rat or a mouse model to study the neural basis of cognition or just cognitive processing in general.

'One of the things that they can help teach us is how animals that engage in lifelong learning do that and why they are capable of doing that when other animals can't. That would actually give us more insight into our own brains and our own cognition.'

Those insights into lifelong learning, Iwaniuk says, could potentially result in new ways of treating or preventing dementia, Alzheimer's, Parkinson's and other neurodegenerative disorders associated with ageing.

But for all the budgie's intelligence and adaptability it can't solve all the challenges that can come with life in captivity.

Chapter 21

Still call Australia home

WHEN THE BUDGIE first appeared in her fairy grotto, Shelley Corvino couldn't have been more surprised had she encountered an actual sprite. Winters in Winnipeg, Canada, are notoriously harsh, and that week of December 2016 temperatures had plunged to overnight lows of –28 degrees Celsius. The Queen of the Fairies herself would have no business outside in the little dell Corvino created among the cedars for the enchantment of her grandchildren, much less a little bird indigenous to Australia's arid zones.

'I really thought I was seeing things at first. We knew right away what he was and that this beautiful lime-green bird did not belong there hanging out with the house sparrows,' the retired Canadian nurse recalls.

The sparrows, fat with their built-in feather puffer jackets, were frequent winter visitors to the Corvinos' house, which backs onto a small lake on the south-east border of Winnipeg, one of the coldest cities on Earth.

Sensibly, then, the sparrows liked to take shelter among the dense clump of tall cedars and, with their usual pickings buried deep in the snow, delighted in the seeds the Corvinos put out. Now here among them was a lurid interloper who not only behaved as if he was one of their number but seemed to speak the same language.

While the budgie was surviving on the fringe of the flock, the Corvinos knew the forecast was not favourable and contacted Avian Welfare Canada, a locally based animal welfare organisation specifically for 'companion parrots'. It was then the Corvinos discovered that the bird had already been at large for more than a month and was troubling the consciences of another couple across the lake, who had spent weeks unsuccessfully trying to catch it before the flock abruptly moved on.

Armed with a heat lamp, a big cage and some helpful advice provided by Melanie Shura of Avian Welfare, the Corvinos applied themselves to the task. Valentino 'Val' Corvino, a born problem-solver who had run an industrial electrical repair business until his retirement, set about improvising a trap in the best style of the cult American television series *MacGyver*. Carefully he rigged up the cage, holding open the door with a stick that could be pulled by a string once the bird ventured inside, drawn in by a cache of its favourite millet seed.

Bravely—and perhaps even crazily, when you are looking down the barrel of bone-chilling overnight temperatures—the couple left a front window slightly open ready to pull the string and trap the bird at any moment. It was to be a long vigil.

At first the cage was on the ground and the Corvinos succeeded in trapping several rabbits in their MacGyvered bird trap. Then, after moving it onto a platform nearer the house and setting up a second heater, they finally got their bird.

'It was New Year's Day we caught him and we were so thrilled,'

Shelley Corvino says. 'The week prior the temperature was dropping drastically. There was a real sense of urgency to get this little fellow out of there. Once he was in the cage he was a bit agitated, but then he saw the big mound of millet in the corner and was very happy.'

The ordeal had left our wee hero, now dubbed MacGyver, with some lung issues. 'For a while it was touch and go and everyone feared we might lose him. He had become community property at this point,' Corvino, who handed him into the expert care of Melanie Shura, laughs.

After MacGyver's remarkable tale of survival was published in the press, no fewer than fifteen people came forward to claim him as their own. Most were genuine; one couple seemed to want a piece of the bird's sudden fame; and one young boy didn't help his chances by telling Shura, 'If you squeeze him and he bites you, he's mine!'

'We live pretty much in the centre of Canada, but I had people calling me from Brandon, from Calgary, from British Columbia, believing this was their budgie. What struck me when I heard their stories about the budgie that got away was the profound sense of guilt and horror they experience and the fact they really needed it to be their budgie,' Shura says.

'One fellow called me from a city 200 kilometres away and left a message saying he was jumping in his car and was on his way. I called him back and got his daughter who said, "That budgie went two years ago. He's dead but Dad just won't let it go." Apparently, he'd been carrying the bird outside to be part of her birthday party a couple of years back when the bottom of the cage dropped out and it took off. He still felt so guilty and wanted to redeem himself.'

It also made Shura realise what great misconceptions people held about their pet budgies and just how many of them were going missing. 'They might have the bird on their shoulder and go to answer the door and then seem surprised when it flies off. The second they are out that door the adrenaline kicks in and in no time the bird is a block away. You can't expect a bird will find its way home by magic when it has never been outside its home before.'

MacGyver's owners were never established, but after Shura lined up a new home for him an extraordinary thing happened. 'We have a flock of birds here and MacGyver fell hard for a lovebird. Lovebirds are kind of bullies and they are powerful, but like budgies they do have an intense need to be bonded and, as soon as he was well, MacGyver bonded almost instantly with a female called Roark.

'He just started feeding her, regurgitating for her, preening her. They do mate or go through the motions. Roark has laid eggs, and of course they are not fertile, but MacGyver regards them as his eggs and watches for them. The pair bonded so it became obvious to me that this was his home.'

Not everyone in the household where birds fly free was happy though. Another lovebird named Pickles snapped one of MacGyver's legs and it had to be amputated. So now he is a one-legged budgie in a cross-species relationship living through Manitoban winters in centrally heated comfort.

'He is very bright and assertive despite his handicap,' Shura says. 'He bosses the cockatiels around, for example, has taken two other rescue budgies under his wing so to speak, and has learned the language of every single species here including the human. He will say in a Larry David voice [David was co-creator of *Seinfeld* and starred in his own show *Curb Your Enthusiasm*] "pretty, pretty, pretty good". He is very communicative, intelligent and I believe capable of empathy. I just find him profound.'

MacGyver is profoundly lucky, certainly. For all its ability to colonise the homes and hearts of people around the world, *Melopsittacus undulatus* remains at essence a true Australian bird which finds itself lost and bewildered when at large, particularly in foreign climes. In the 180 years since it was first successfully shipped off to England, the budgie has proved that you can take the bird out of Australia, but you simply cannot set it free with any reasonable prospect of survival.

Where other introduced birds such as rose-ringed parakeets and monk parakeets have thrived in feral colonies outside their natural habitats, the budgie has effectively forgotten how to, and has dismally failed to establish itself beyond its native land in any sustainable numbers.

Generation upon generation of captive and selective breeding has produced, at best, a much-loved and cosseted companion and, at worst, a feathered Frankenstein that's not so much a bird as a caricature of the original wild creature.

 ## Where budgerigars abound

There is no grander site in Australia than a vast plain, strongly grassed in spring, spreading to the very horizon, with a running wind scudding it in waves . . .

It was perhaps half an hour before sundown when we first saw a black cloud far to northward, just above the horizon. Because of the cloudless sky and the cool evening, this cloud was very definite, jet black in the distance at first, moving eastward. It rose and fell, with a long winding tail. Then it turned and flashed with myriad points bright yellow in the sun, swirled up and over, and lashed eastward again like the crack of a long whip. Seconds later, the black cloud swirled like the water of a whirlpool, and this time flashed myriad green.

Breathlessly, we stopped the car and got out into the thick grass. Another dark cloud had swept into view across the [western Queensland] horizon, larger than the first; and yet another. They were coming from the north-west. Budgerigars across the plains; budgerigars in tens of thousands; small birds flying fast in pack formation; tens of thousands of bright green and yellow, chattering feathered rockets, each mob at some common impulse wheeling and turning as one; spiralling, twisting, darting, climbing up and over in a grand loop to catch again the rays of the setting sun; spreading feet above the grass. Then they appeared as one with the plain, racing across its flat surface; sharply upward again, in joyous climb a hundred and two hundred feet, to the level down and race again.

It was surely a happy omen, silent and distant at first. We did not know which way to turn; they crossed the horizon in vast mobs— perhaps 20 or 30 mobs, until millions of birds were racing in mass excitement. Our excited voices were drowned by the sound. It had grown speedy, terrific with its flash and rumble; disconcerting perhaps. Human eyes could not follow the movements of more than two or three flights at a time.

We drove through the thick grass, slowly, to the bank of the creek, wide and sandy, with water still fresh in pockets at the swirled-out roots of coolabahs. By now the whir of wings had become continuous. The silent land of an hour ago was alive with sound and amazing movement. The birds were shooting like green bullets through trees and bushes, up and down the river-bed.

Budgerigar

They were oblivious to human presence. They were speeding for life, with hawks ready to pounce. The battle was apparent. Fifty and 100 feet above the tree-tops, gliding hawks circled in hundreds; and almost as the sun went down they began to dive. Many of the dives were close misses; but now and then there would be a louder, startled rumble of wings, momentarily, as thousands of budgerigars swerved instinctively, leaving one of their number screeching and biting at crushing claws.

Small green, yellow or white under-feathers floated in a steady drift to the sand. It seemed incredible that small birds could fly so long, so fast, so alert to danger. As dusk deepened, the rumble of wings became intermittent. Perhaps five or ten thousand birds would swerve suddenly from the centre of the creek to vanish, like the shutting off of light, into the branches of trees; too quick for human eyes to follow . . .

It was not until an hour after dark that the creek was silent. We knew that millions of birds had gone to rest, waiting for daylight. Was it five million? Ten million? No one will ever know. By torchlight we approached and saw them perched and hanging in dozens to a leafy frond, hundreds to a large branch, thousands to a tree . . .

In the morning light, these millions of birds had gone out from the creek, both north and south; therefore I could only guess they had come by mass instinct, many hundreds of miles from all the parched lands across drought-stricken Northern Australia, leaving behind only scattered dozens.

They had found, for the time being, a fertile place. Drought, to them, was part of a great cycle, starting yesterday, ending tomorrow. They knew all the answers.

—Arthur Groom, conservationist and author (1904–53),

Wild Life, June 1953

It's not simply a matter of habitat and climate. The odds of an escaped cage bird's survival in Brisbane are almost as grim as it might be in Budapest, as Professor Darryl Jones, urban ecology expert and author of *The Birds at My Table*, explains.

'If they have been in cages for more than a few generations there is no way they can cope. That is especially the case with the social species of parrots which really do need to be in groups to learn how to do things and breed,' Jones, the straight-shooting Griffith University professor, says.

'Wild budgies scream around the sky in huge numbers and fly massive distances. Cage budgies are up to around three times the size and they are not very fit. They aren't set up to be successful when they have left cages, especially if they have been umpteen generations in captivity. They don't last long.'

Jones has personally witnessed examples of budgerigars trying to keep up with rainbow lorikeet flocks. 'They kind of figured out they were like parroty-things and hung out with them,' he marvels.

'You would see the rainbows flying past and then 10 metres behind would be this tiny budgie trying to keep up. I saw one that managed to stay with them for a few months, then they all moved on and I don't know what happened. They just kind of ignored him, but he would be on the ground foraging around with them. He was a classic blue so stood out a mile and more than likely he would get picked off by a predator at some stage.'

Budgies on the lam are at a definite disadvantage because of what people have done to them over many decades of selective breeding.

'They have been through a definite bottleneck in terms of genetics and the domesticated ones are all a bit inbred, though some still have a bit of capacity, so they are hanging around near someone who is feeding rosellas and lorikeets,' Jones says. 'They would love to be back in their cage if they could find it again, but they have gone too far and don't know where they are anymore. They are reliant on the feeders. You are never going to have rabid groups of feral budgies descending on you.'

This, however, was exactly what authorities feared in Florida in the 1970s. The Florida Game Commission conducted several studies into the populations of introduced birds, including the budgerigar, amid fears they could pose a threat to the viability of indigenous species, particularly those like the northern flicker that also build their nests in tree cavities.

Budgies, both escaped and deliberately released to be fostered as an attraction, have been recorded across 67 counties of Florida since the 1930s, most prevalently on the coast of the Gulf of Mexico from Pasco to Sarasota. In the late 1970s the feral Florida population was estimated as being about 20,000 strong, with as many as half of these concentrated in Pasco and Pinellas counties. They were reported nesting in natural pine-tree cavities, artificial nest boxes, cabbage palms and even streetlights in areas where there were bird feeders.

Then the population collapsed without any obvious reason, although several particularly cold winters and intensified competition for nest sites from the burgeoning population of European starlings are certain to have contributed. By the 1990s the Florida population was believed to have dropped to around 200, with budgie sightings continuing to decline in recent years according to the Audubon's Christmas Bird Count. Even so, it is still among the most common parrots found as escapees by atlasers.

It was a similar story of boom and bust with a small colony established on Tresco, the second-largest of Cornwall's surprisingly temperate Scilly Isles. In 1969 and 1970 a total of ten pairs of free-flying budgerigars—offspring of the royal flock at Windsor—were introduced to the gardens of Tresco Abbey by members of the Dorrien-Smith family, whose ancestors had once held the title of Lord Proprietor of the Isles of Scilly.

The budgies were thought to be the perfect addition to the family's Tresco Abbey estate, and for a while it seemed the birds were as at home as the golden pheasants that wandered the grounds. By 1974 there were at least 35 nesting pairs and a 100-strong flock that eschewed the nesting boxes of the open aviary and happily settled into holes in the elms, sycamores, cordylines and palms. Soon they were also to be seen in a number of the smaller neighbouring islands, much to the dismay of ornithologists who were keen to protect the already decimated numbers of migratory puffins, Manx shearwaters and sparrow-sized storm petrels from more foreign invaders. However, when one of the elder Dorrien-Smith women, who had been supplementary-feeding the budgies, left the island, the population crashed.

Australian backpackers may be notorious for their ubiquitousness but they have nothing on the great diaspora of *Melopsittacus*, who can be found in some very surprising contexts. The award for the most intrepid budgie goes to the bird discovered on an oil rig almost 20 kilometres off the coast of Caithness, northern Scotland, and flown by helicopter to the safety of Aberdeen.

Chapter 22

Sooths and sayers

IF EVER THERE was an auspicious time for a hardy, budget parrot with the ability to mimic human speech to be let loose on the world, it was the second half of the nineteenth century. Some might even take it as an augur that budgiemania coincided with an explosion in spiritualism and dovetailed so nicely with the Victorians' fascination with all things occult.

Communing with the dead had become a preoccupation in the English-speaking world since 1848, when two young American sisters, Kate and Maggie Fox, claimed to have made friends with the spirit of a murdered peddler inexplicably buried in the cellar of their family's house in Hydesville, New York. Gazing into crystal balls, and reading palms, tea leaves and cards flourished among all levels of a society seeking some supernatural comeback to the ascension of science. From the tittering innocence of parlour games divining the name of young women's future husbands to the full theatre of seances with animated furniture and the appearance of radiant ghosts in darkened rooms, people were seized with a passion for the paranormal.

In 1824, the Vagrancy Act in Britain declared that 'every Person pretending or professing to tell Fortunes, or using any subtle Craft, Means, or Device, by Palmistry or otherwise, to deceive and impose on any of His Majesty's Subjects' would be deemed 'a Rogue and Vagabond' and convicted of an offence. Yet these practices flourished, in part because the spiritualist movement had many highly placed supporters.

The wife of US president Abraham Lincoln became convinced she could commune with their son Willie, who had died at age eleven of typhoid fever. Queen Victoria and Prince Albert, too, had attended seances together, and after her beloved husband died in 1861 it is claimed she believed a thirteen-year-old Leicester boy, Robert James Lees, channelled Albert's spirit, and she invited him to attend seances on numerous occasions. Sir Arthur Conan Doyle was an avowed spiritualist who created the Hound of the Baskervilles and Sherlock Holmes as figures of fiction while literally believing in fairies at the bottom of the garden.

Suspended prettily in its cage against this backdrop, the budgie would soon be co-opted as 'The Fortune-Telling Bird'.

Since ancient times, humans have interpreted the movements of birds as general omens of the future in a form of divination known as augury or, more specifically, ornithomancy. In Victorian London this was taken to the next level, with individual birds trained to leave their cages upon a command and appear to carefully assess a customer, before selecting a card or paper from a box upon which was written the customer's fortune.

Lucky Joey

HOVE, England—Joey, a chirpy three-month-old pet budgerigar, has been doing soccer pools for only a few weeks but has already pecked out nearly £500 in prize money for his owner Ernest Mahoney.

Mahoney writes down the numbers of the soccer teams on pieces of paper. Joey pecks here and there, comes up with some of the pieces on his beak, and his owner marks the numbers down on his coupon.

Last week Joey netted more than £400 in one pool. The week before he came up with a small fourth prize.

—*The StarPhoenix* (Saskatoon, Canada), 20 October 1954

Even in the 1880s, parrot astrology was hardly new, with the practice of *kili josiyam* originating in the southern Indian states of present-day Tamil Nadu and Kerala and spreading to Singapore. The Indian fortune tellers traditionally used rose-ringed parakeets, indigenous to the subcontinent, to select one of 27 cards representing the nakshatras or 'lunar mansions' of Vedic astrology. In 1972 it became illegal in India to cage any native birds—including rose-ringed parakeets—under the Wildlife Protection Act. This, coupled with a growing awareness of animal welfare, has had the effect of putting parrot astrologers out of work. However, practitioners and their birds can still be found along Serangoon Road in Singapore's Little India—the most famous of these being M. Muniyappan and his assistant Mani the parakeet, who became much sought after for picking the correct winners of all four quarterfinals of the 2010 World Cup. Mani was knocked off his perch in the finals by the superior powers of Paul the Octopus, an oracle from Germany.

Today, at the entrance to the exquisite Musalla Gardens in Shiraz, Iran, visitors to the marble Tomb of Hafez seldom pass up the opportunity to engage in a more lyrical version of birdy prophesy.

A finch or budgerigar will pluck a verse or quotation by the great fourteenth-century poet from a box.

The British version involved budgies and 'Italians', who were quite a familiar part of the street trade in London in the late nine-teenth and early twentieth centuries. In 1886 one of their number made the news when she was one of five people, including two ten-year-old boys, crushed to death when the front walls of four London houses that were in the process of being demolished were blown down by a sudden strong wind.

 Fall of four houses in London

Five persons buried

. . . The catastrophe was witnessed by a large number of persons. No one was certain whether anybody had been buried in the debris or not, but communications were immediately sent to the neighbouring police-station, and a large force of police were soon set to work clearing the spot of people and assisting a body of men dispatched by the Islington Vestry, in searching the mass of brickwork and rubbish.

In a very few moments the body of an elderly Italian woman, fearfully crushed and covered with blood and dirt, was brought forth to the horror of bystanders. It would have been difficult to recognise the unfortunate woman, but for her Italian attire. She was a well-known figure in the Holloway-road, having for a long time occupied a stand outside the fallen buildings with her case of birds which were 'warranted to tell your true fortune'. One of the birds escaped, but the others are believed to have been killed.

A touching scene occurred sometime afterwards when Antonio Giuseppe Forto, an organ grinder, recognised the body at the

Chapel of Ease Mortuary as his wife Francesca. Both he and his son were quite broken down with grief and had to be helped through the streets by their friends. The deceased was 50 years of age.

The fifth body recovered was that of Mrs Caroline Taylor, aged 45. The deceased had been attending to her daughter, who was in a dangerous state of health, and was on the way home from a fish shop with some soles for the patient when she met her untimely end. When she was discovered she was still grasping fish in her hand.

—*Central Somerset Gazette*, 6 February 1886

The curious thing about the use of budgerigars by street entertainers and hawkers in this period is that the birds said nary a word. They had been trained not to talk but rather to listen to the commands of the handler, which they did more reliably than mimic set phrases of human speech.

Chapter 23

Famous fanciers

WITH ONE HUGE hand Geoff Capes deftly plucks a pied budgerigar from the flight, pulling it upside down towards him, then flips his wrist to reveal it lying motionless on its back in his palm. Gently he strokes the creature's legs down with a single meaty fingertip. 'There's a tame one for you,' Capes tells astonished visitors who have come to his aviary in Lincolnshire, UK, with some quite unrealistic ideas about what £15 might buy. Seeking to validate his own eyes, one man tentatively reaches out to touch the seemingly insensible bird, which immediately rallies and bites him for his trouble. 'Just a little trick I have learnt. Works every time,' Capes chortles.

While age and a degenerative back condition mean he no longer quite stands at 197 centimetres, Capes is still a towering celebrity. Twice crowned the World's Strongest Man, capable of running 100 yards with a 400 pound fridge strapped to his back and of pulling a 165-ton two-carriage train, Capes was a household name in the UK throughout the 1980s and even inspired a rudimentary computer game called *Geoff Capes Strongman*.

As the most capped British male athlete of all time, Capes twice won Commonwealth Games gold in the shot-put and represented his country 67 times, including in three Olympics. He still holds the record as the most successful Highland Games competitor, winning the championship six times and earning the nickname Geoff Dubh Laidir, Scottish Gaelic for 'Black Strong Geoff', after his dark hair and beard.

Grey-haired now at 70, and struggling with health problems, Capes still remains highly competitive in the unlikeliest arena—the budgerigar show scene, where he is highly respected as a champion breeder, judge and former president of the Budgerigar Society of Great Britain. The most famous recipient of Capes' birds was Yul Brynner, the Academy Award-winning actor (*The King and I*), who was known for breeding Roller racing pigeons and keeping penguins. One can well imagine Brynner raising an elegantly arched eyebrow in bemusement at the decorative little poppet.

Capes was first introduced to the birds as a young lad by his father, who used to breed them in a repurposed wardrobe. 'When you opened the wardrobe there was a wire front with nest boxes inside and holes in the side where the birds could come out into a little flight,' Capes recalls.

It was not until he was nineteen and in his early days on the beat as a policeman that he became interested in the hobby himself. 'I went to arrest a guy. It was not a serious matter, just a non-payment of fine,' Capes says. 'Knocking on the front door I could see he had all these budgerigars in the front room. He eventually came to the door and I asked if I could have a look at his budgerigars. I went in and he showed me all the different colours and everything. I was there for about an hour and ended up by saying, "Oh, and by the way, you are under arrest", and served my warrant. I went back about a week later and got three pair of budgerigars off him. There were no hard feelings.'

Capes kept budgerigars all throughout his sporting career and regards it as a kind of therapy, a release away from his sporting and strongman competitions. 'I had a long career and they were a whole part of it, going into the bird room and just sitting there. Looking at the progeny and visually pairing them up in your mind. Looking and thinking, that one's got a better face than that and that's got a longer body and would be good to pair with that.'

A release from tension was necessary in the highly supercharged environment of elite athletics. This became painfully apparent when Capes was disqualified from the 1978 European Championships in Prague for shoving an official who tried to stop him as he approached the shot-put circle in the final because he was wearing one number on his vest instead of the regulation two.

Long retired from the sport, Capes uses those same hands that once hauled a fully laden truck across a field to gently wash, blow-dry and clip the nails of his best birds in preparation for the latest show. 'I like showing, which is semi-competitive, and being a competitive person, I like to do well. Recessive pieds are my big strong section, but I have done well in others. In 2014 I got best young cinnamon and best cinnamon in the World Championship, and I have won many, many championships with my normal and recessives.'

Capes is far from the only high-profile athlete to keep birds as a counter to adrenaline-fuelled and testosterone-charged sports. James Hunt, the 1976 Formula One World Champion, was infamous for his love of women with the words 'Sex, breakfast of champions' embroidered on his racing overalls.

Hunt was rather more steadfast with his budgerigars, with a collection of between 250 and 300 birds at his home in Wimbledon, London. He was a regular on the show scene and was said to prize the rosettes and ribbons he collected for his birds as highly as his racing trophies. After Hunt lost his fortune, and gave up smoking and drinking, he would spend hours in his aviary and happily talked about the birds given any opportunity.

The budgerigars found their way into the 2013 biographical film *Rush*, based on the story of the rivalry and friendship between Hunt and Austrian driver Niki Lauda. In one scene Hunt, played by Australian actor Chris Hemsworth, declares: 'I fall out with people left and right. The only creatures I have every really loved or treated honourably are budgerigars.' It's a line that possibly suffers from lack of context.

It does seem that these small birds have an astonishing capacity to bring out tenderness in great and powerful men. Toby, the afore-mentioned pet of Sir Winston Churchill, was indulged in every respect and privy to the most important discussions of budget and policy among the prime minister and senior members of his cabinet.

At breakfast Toby would be there challenging his own reflection in the silverware or parading up and down the damask linen table-cloth with a tiny spoon from the salt cellar—a favourite trick. When Churchill was writing his memoirs or letters Toby would perch on his master's glasses and sometimes on his head.

Writing to his wife Clementine on 8 August 1955, Churchill acknowledges Toby as co-author:

My darling, I was so glad to get your letter . . . Now it has come I take up my pen to answer, aided by Toby, who is sitting on the sheet of notepaper insisting on lapping the ink from my pen in order to send you a personal message . . . He is a wonderful little bird. He pecked and scribbled with his beak and what I have written so far is as much his work as mine.

At this stage Toby, a blue and green cock who had been given to Churchill by his son-in-law Christopher Soames, was not yet a year old but he became so ubiquitous that the Chancellor of the Exchequer Richard 'Rab' Butler carried a special silk handkerchief to wipe the bird's droppings from the top of his balding head. Leaving an audience with the PM on one occasion besprinkled with budgie poo, Butler sighed, 'The things I do for England.'

Toby travelled everywhere with Churchill, including to his frequent retirement sojourns to the Riviera where the bird was treated to rosewater baths in a silver bowl in a luxury white marble villa designed by and built for Coco Chanel.

He was also a frequent seafarer, joining Churchill aboard Greek shipping magnate Aristotle Onassis's 100-metre-long superyacht *Christina O*. It was during one of his visits to the Côte d'Azur in 1961 that Toby flew out the window of Monte Carlo's Hôtel de Paris penthouse suite and was last seen in the vicinity of the famed casino.

Churchill was greatly upset, and Onassis took it personally that his friend's bird should go missing on the first afternoon of what was to be a two-week holiday. The billionaire mobilised hotel staff, police, his own private security team and the fire brigade and stayed up all night to conduct the search. Churchill offered a reward of 150 francs to no avail. A bird matching Toby's description found on the road near Nice 20 kilometres away was declared to be an imposter by Churchill's personal detective, who drove down to examine it.

Onassis gave Churchill a replacement budgie named Byron which proved somewhat irascible and was sent to Field-Marshal Bernard Montgomery, also a well-known budgie breeder, in hope he could curb its biting.

Whether it was this bird rehabilitated or another altogether is unclear, but there was a budgerigar beside Churchill right up until the final months of his life. Notes written in 'The Form', a notebook containing a daily record of events by nurses involved in the care of Churchill up to his death on 24 January 1965, reveal 'the bird' was high on the list of their duties. 'Whisky and soda, specs, cards; bird to be brought into dining room near his chair' says one entry. 'After dinner put bird to bed!' advises another.

 Famous budgies: DISCO

Confirmation of the death in January 2017 of Disco the Parakeet prompted an extraordinary outpouring on social media. As the first true budgerigar internet superstar, with his own YouTube channel and videos viewed over 19 million times, Disco was compulsory viewing for many people. Some even confessed to having Disco sayings as ringtones on their phones.

Rumours that Disco died in a horrific accident began to circulate after several months without updates to his Facebook page. Acknowledging the death of the 40-gram celebrity, his distraught family revealed the truth was more prosaic. They had simply found the 6½-year-old's lifeless body at the bottom of his cage one morning—cause of death unknown. The death notice attracted an extraordinary 10,000 likes, 3700 consoling comments and 1300 shares.

Disco began talking within weeks of going to live with Judy and Kevin Bolton and their daughter Addie in Rochester, New York, demonstrating a remarkable ability to mimic just about anything from Monty Python to Rolling Stones classics. According to his Facebook blurb he could 'beatbox, snore, bark and meows better than some cats'. His principal personal interests were listed as millet, paper aeroplanes, world peace, yodelling, coriander and red lettuce. He even learned some Swedish to talk to one of his feathered fan base in Malmö.

To disprove detractors who believed the bird was being dubbed in videos, Disco appeared on NBC's *Today Show* and also featured in the two-part BBC TV documentary *Pets—Wild at Heart*.

He remains sadly missed.

Surprisingly, budgerigars also played a bit part in the early life of the one of the world's great spiritual leaders. After the fourteenth Dalai Lama was enthroned in 1939 in Lhasa, Tibet, emissaries from the leaders of other countries arrived laden with gifts for the then-four-year-old. The Maharaja of neighbouring Sikkim, for example, furnished the child with offerings including a pair of sporting rifles; a tiger skin over 3 metres long, complete with tail; and a pair of English horses, which were procured with some difficulty but gratifyingly towered over Tibet's hardy mountain ponies.

Budgerigar

The English, represented in the form of yet-to-be-knighted Sir Basil Gould, the British Political Officer for Sikkim, Bhutan and Tibet, were not to be outdone. Gould delivered three rifles and an English horse saddle, plus a freshly minted bar of gold, ten bags of silver, rolls of cloth, a gold watch and chain, a picnic basket, a music box, a garden hammock, a cuckoo clock, a pedal car and two pairs of budgerigars.

As the little lad, born in a cowshed but now swathed in brocaded robes, was reverentially lifted onto the 2-metre-tall golden throne in Potala Palace, the offerings piled up in an anteroom. The budgerigars, having survived the long journey on the back of pack mules through the Himalayas in below-zero temperatures, were included in the list of gifts but sensibly spared the long ceremony and placed in the care of the British Mission's wireless operator Reggie Fox.

His Holiness was then as wilful as any young child and very fond of birds. When he heard of the budgies, Tenzin Gyatso—as he had been renamed upon entering monastic life—wanted them and he wanted them now. 'There came a messenger from the Potala to request immediate delivery of the birds; then two more messengers, more senior than the last; and then two more,' Gould later recalled.

'It was soon clear that, if there were to be a battle of wills, the Dalai Lama would prove that his will was the stronger; so it was decided that compliance was the only possible course.' A clerk, Pemba Tsering, was dispatched to the palace with the birds: 'Pemba, considerably overcome, handed over the birds, and tried to make himself scarce, but he was sent for by the Dalai Lama who, talking Tibetan clearly and easily, discussed the birds' food and how to keep them safe.

'And there was evidence of the Dalai Lama's real kindness to animals when a few days later, being persuaded that they might be

206

better off for the time being in Mr. Fox's kindly care, he sent the budgerigars back to Dekyi Lingka [to the British Mission], where they became great favourites with visitors.'

From the saintly to the rather less so ... famed British conductor and composer Sir Malcolm Sargent killed his first budgerigar with an excess of sherry, but was more careful with his beloved Hughie who used to sit atop his head when he was having a bath. In fact, until fairly recently, sherry or cough mixture was added to the budgie bathwater by some serious bird exhibitors because the sweetness encouraged birds to preen their feathers more, so they looked particularly spruce on the show bench.

Continuing the alcohol association, the husky-voiced screen siren Tallulah Bankhead had a champagne-quaffing parakeet called Gaylord, a habit that possibly helped the little bird to cope with the constant sidestream from her mistress's 100-a-day cigarette habit.

In 1959 Hollywood siren Jayne Mansfield, a known bird lover, astoundingly accepted an invitation to judge a budgerigar show at a church hall in the heart of London's notorious East End. Mansfield, who was living in the UK during the filming of the gangster thriller *Too Hot to Handle*, found herself closer to the real deal than expected when she handed her coat to a teenage bystander while she posed obligingly for admirers and local press. The teenager was Tony Lambrianou, who later became infamous for his association with real life gangsters the Kray twins and was sentenced to fifteen years for the murder of another gangster. As legend has it, Lambrianou made off with Mansfield's coat and flogged it.

The world's most enduring sex symbol, Marilyn Monroe, kept two parakeets, Butch and Bobo, which she taught to talk. During her marriage to playwright Arthur Miller the birds frequently accompanied Monroe as she shuttled between their ranch in Connecticut, New York and Los Angeles. Butch, the more voluble of the parakeets, was keen everyone should know where his heart lay and to the delight of fellow passengers would often announce, 'I'm Marilyn's bird, I'm Marilyn's bird.'

In that mid-century period later mythologised as 'Camelot', there were also two budgies named Bluebell and Marybelle living in the White House. The birds were pets of the Kennedy children. When Jackie, Caroline and John Jr left the presidential residence and workplace following the assassination of JFK, the birds were the first members of the family to be installed in their new home in Georgetown, Washington DC.

Another budgie keeper was poet and novelist Vita Sackville-West, who kept a flight of more than 30 budgerigars in her gardens at Sissinghurst Castle in Kent. Budgerigars flit through the last letter sent to Sackville-West by her former lover, Virginia Woolf, in March 1941. There was no inkling she was about to take her own life in the fond epistle, in which Woolf talks of her cook and housekeeper Louie Mayer's budgerigars surviving the war on scraps, and flagging the prospect of bringing some back for her from Sackville-West's aviary. 'I suppose they're lower class, humble, birds. If we come over [to Sissinghurst], may I bring her a pair if any survive? Do they die all in an instant? When shall we come? Lord knows—,' Woolf wrote on 22 March 1941. Six days later she filled the pockets of her heaviest fur coat with stones and walked into the Ouse River, Sussex, to her death.

Budgerigars and children's authors can be paired in happier tales. Beatrix Potter counted one among her family menagerie growing up and although none featured alongside Benjamin Bunny, Peter Rabbit and the other animals in her works, the Victoria and Albert Museum in London holds a watercolour and pencil sketch of a budgie by the illustrator. Roald Dahl, that giant of children's literature, delighted in his flock of up to 100 homing budgerigars, which darted freely among the trees of Gipsy House in Great Missenden, Buckinghamshire.

A blue budgerigar named Dempsey became the constant companion of the great British novelist, travel writer and war correspondent Maurice Baring when he was confined to bed by Parkinson's disease and unable to write.

When Sir Billy Butlin, the holiday camp entrepreneur, famously quit Britain for the Channel Islands for tax reasons, he created a haven in Jersey that was home to scores of peacocks and 600 free-ranging budgerigars.

As a kid, Australian novelist Robert Drewe bred budgies with a slightly more mercenary mindset. He raised one particularly precocious budgerigar called Junior that he hoped might win the BBC's talking-bird contest famously won by Sparkie Williams. Junior, who lived for more than fourteen years, had a repertoire including 'The Banana Boat Song' and 'Tie Me Kangaroo Down, Sport', but the folk at the BBC never called in response to the young Drewe's tape-recording.

Another young fancier who made her mark on the world was Australia's Golden Girl and Olympic champion Betty Cuthbert, who kept an aviary of birds she bred and sold in her spare time, in between working in her father's nursery and training.

Budgerigar

 Australian lassie is for the birds

Betty Cuthbert, sprint champion raises budgerigars

MELBOURNE—The new world sprint queen is a comely Australian teen-ager whose chief hobby is raising Budgerigars—talking love-birds similar to parakeets in America.

'They are my only love outside of running,' said Betty Cuthbert, 18, after adding the women's 200-meter running crown in record matching time today to the Olympic 100-meter title won earlier. 'I breed them and sell them but keep one of my own—a pet—which I wouldn't sell for the world.'

Betty is an assistant to her father, a nurseryman who lives 12 miles outside Sydney.

She said she had no particular strategy in today's race. But was told by her coach to hug the line and try and get into the lead before reaching the straightaway.

Her 23.4 seconds clocking matched the world and Olympic records.

—*The Record* (New Jersey), 30 November 1956

And, proving it is never too late to be bitten by the budgie bug, Ford Rainey—the American actor who appeared in *The Sand Pebbles* with Steve McQueen and had recurring roles in a raft of television shows including *Bonanza, Gunsmoke, The Bionic Woman* and *The King of Queens*—took up breeding and showing budgerigars at age 90. By the time of his death at 96 he had won scores of trophies and ribbons for his budgerigars in shows around southern California.

Meanwhile, in a more contemporary example of budgies' Beverly Hills celebrity appeal, singers Gwen Stefani and Britney Spears both bought parakeets for their children. The well-known actor and producer Aston Kutcher once gave budgies to his wife, actor Mila Kunis, as a Valentine's Day gift, believing them to be lovebirds—a term once ascribed to budgerigars but now firmly bestowed on the *Agapornis* genus of parrots that are truly monogamous. It was a satisfying example of the *Punk'd* star being, well, punked in a case of dodgy bird identification. 'We took them to the bird lady . . . she informed us that we, in fact, did not have lovebirds, but instead have parakeets. They're called, like, bungees? We have bungees,' Kunis laughingly told talk-show host Ellen DeGeneres.

 ## Famous budgies: GUINNESS WORLD RECORD HOLDERS

When it comes to talking, no bird has yet equalled the record held by a perky blue American parakeet named Puck Jordan. Puck was accepted for the 1995 *Guinness Book of World Records* only a few months before he died of testicular cancer. The bird, owned by Camille Jordan of Petaluma, California, was acknowledged as having 1728 distinct words verified by 21 volunteers over six months and backed up by 30 hours of tape recordings and videos by his owner.

By comparison Oskar, a German bird accepted into the 2010 *Guinness World Records* for having the largest vocabulary of a then living bird, seems positively mute. Oskar was found to have just 148 words when tested in a single session. He could also speak 50 meaningful sentences and was bilingual, having been taught in both German and Polish by his nurse-owner, Gabriela Danisch.

The record for the longest-lived budgerigar is still held by Charlie Dinsey, who lived for an astonishing 29 years and 60 days in the care of one J. Dinsey of Stonebridge, London.

Chapter 24

Spinning around

IF IT WAS remarkable that one little Australian bird could gain world ascendency, in 1988 it was the turn of quite another. Kylie Minogue, already a star in Australia and the UK thanks to her role as overall-wearing mechanic Charlene in the television soap opera *Neighbours,* flew to the top of the pop charts in ten countries with her second single 'I Should Be So Lucky'.

The song, which had been written in 40 minutes by the hit-churning British partnership of Mike Stock, Matt Aitken and Pete Waterman and recorded by Minogue in less than an hour, became one of the great earworms of our time. There was no escape from its bright, boppy catchiness. Market research found that the song so annoyed people in the age 25–40 cohort that several radio stations briefly banned it from their playlists.

Even the broadsheets could not ignore the pint-sized phenom-enon that was Kylie, so it was with tongue in cheek that Anthony Dennis, editor of *The Sydney Morning Herald*'s 'Today's People' column, referred to La Minogue as 'The Singing Budgie'.

'It was a light-hearted item where I suggested if France could have a "Little Sparrow" [the nickname given to legendary and petite French chanteuse Edith Piaf], then surely Australia should have a Singing Budgie,' recalls Dennis, now Fairfax Media's national travel editor.

Dennis also takes credit for coining the phrase 'The Human Headline'—worn with pride by radio and television host Derryn Hinch—and insists the budgie nickname was not meant as a slight. 'It was originally intended as a bit of whimsy, and at the time she was utterly ubiquitous.'

The sobriquet stuck. While it was rumoured that Minogue did not like it and felt it belittled her, it seems that this was an assumption, and that actually she had taken ownership of brand budgie, as she'd done in all other aspects of her career.

On 29 January 1990—when Minogue had already made a quantum image leap from saccharine poppet to kinda cool with the help of her then-boyfriend, the charismatic INXS rocker Michael Hutchence—she played a secret warm-up show to her Enjoy Yourself Tour in her hometown of Melbourne. In what could only be regarded as flipping the bird to naysayers who doubted she could pull off a full-length solo show, Minogue and her band took to the stage of the less-than-salubrious Cadillac Bar in Swanston Street that night under the name The Singing Budgies.

Dennis couldn't have come up with a better nickname for the singer, who has proved as hardy and adaptable on the international stage as *Melopsittacus*, repeatedly bouncing back just when it seemed her star had waned. The birds, too, have been alternately mocked, merchandised, mythologised and subjected to the vagaries of fashion. But they always put on a good performance.

Memorable occurrences among budgerigars

Like most nature-lovers I have witnessed many memorable incidents among birds and other creatures; but I think the night singing of budgerigars will always be my loveliest memory.

With farmers and gardeners, birds must have been sighing for rain during the unusually long dry spell of May sunshine this year. At 7 p.m. on May 23 rain fell and our budgerigars appeared to rejoice. During a lull in the heavy downpour my daughter called to me to hear them singing while the rain was still pattering. She had heard them above the sound of heavy rain, with the wireless on, through a closed window some 25 ft. distant from the birds. We listened for a little while through the open window, then went out to the aviaries.

It was quite dark but the 'budgies' sang on. The usual sounds that call us out to the birds are cries of alarm or a noisy fluttering when something has frightened them; but this was—how different! The united voices of some 130 comfortably sheltered birds were singing in the rain. The sweet, clear notes suggested the ripple and gurgle of water; or the patter of the rain itself—a full chorus rejoicing over the breaking of a Maytime 'drought'.

One has no words to convey adequately the wonder and beauty of that night chorus. We could not have spoken, had we wished, as we stood there in the dark, listening to Australia's cheeriest birds singing their song of the rain. It brought a fuller realization of what rain must mean, under natural conditions, to these merry little birds. It was eloquent of their rejoicing at the end of a long dry spell, or the breaking up of a real drought, and the springing and seeding of the grasses they love.

What a difference in the air they must have sensed to call forth such a paean of praise! It is almost pathetic to note their delight in freshly-gathered moist grasses.

Something of what rain must mean to them may be surmised from their joy when a garden spray is directed into their houses. The sound of the sprinkler draws them fluttering hopefully to the wires. They dash in and out with open, beating wings until they are soaked.

A clump of wet grass, taken with a spit of earth, fills them with ecstasy. They tumble through the green blades until not a drop of moisture can be left on them. A bunch of cut grass is treated the same way until the whole is scattered.

Like other birds in natural conditions budgerigars would bathe by darting in and out of dew and rain-wet branches. This was revealed in their instinctive knowledge of what to do with wet gum-twigs hung on wires—cage-bred birds which had never seen or felt a dew-laden tree!

—Edith Coleman (1874–1951), naturalist and writer,
The Victorian Naturalist, 1947

It was not for nothing that John Gould had named them *Melopsittacus*, the melodious parrot. The budgies' own discography is also quite considerable. The first commercial recording of a budgie was a 78 rpm of 'Joey the Budgie' produced by Queensway Recording Studios in London in 1952. This was followed by a double-header, released by Decca in 1955, featuring Bradford's Beauty Metcalfe which had a thick Yorkshire accent and Sandy Paul, a yellow bird from Middlesex. The best-selling Sparkie Williams disc came three years later, and in 1962 came its spin-off *Pretty Talk*, which consisted of Sparkie's owner repeating simple phrases to demonstrate how owners might also train their birds to speak. Philip Marsden, an authority on the birds, gave tips on them in *Talking Budgerigars*, put out in 1967. Also getting in on the act was entertainer Bruce Forsyth, who recorded the chirpy song 'My Little Budgie' with 'I'm a Good Boy' on the flip side in 1960. Across the Atlantic, the label Hartz Mountain produced a record by which 'your parakeet can train itself to talk' but failed to explain how the bird might place the *Parakeet Training Record* on the turntable.

But budgies are incredibly smart. Recent studies have shown that these little birds can pick out abstract patterns in speech, making them the first non-human species to grasp simple grammar. Research has also demonstrated that they can count to six, recognising which closed containers contain food by the number of dots painted on the top of them. They also yawn contagiously, just as humans do—a trait linked to empathy and previously only found in other primates and dogs.

Budgies also have a sense of rhythm and can dance to music like other parrots, although admittedly not with the same proficiency as Snowball the Cockatoo, who became an internet sensation in 2007 when he was filmed busting moves to the Backstreet Boys hit 'Everybody'.

The budgie has been painted by masters, coveted by royals, rendered in the finest porcelain, modelled in plastic, dipped in batter, baked in pies, bred artificially in test tubes and depicted on the postage stamps of more than 30 nations from Antigua and Barbuda to Zambia. In 2013, it was even represented on a half-ounce silver proof coin by the Perth Mint.

Budgies are embedded in the lingo of their home country. A 'dunny budgie' is a giant blowfly, while 'budgie balls' reflects quite the opposite view of testicles, particularly those belonging to men on steroids. In World War II the word budgerigar was used to describe soldiers who frequently wrote to wives or girlfriends. If a tent had 'a number of budgerigars it became an aviary', according to an article on army slang published in *Smith's Weekly* in 1944.

The birds are identifiable in major languages throughout the world. As a sampler, the budgie is *undulaatti* in Finnish, *Wellensittich* in German, *volnistyy popugay* in Russian to *xiǎo chángwěiyīngwǔ* in Chinese and *chim vẹt đuôi dài ở Úc* in Vietnamese, while the Turkish know it as *muhabbetkuşu*.

There have been dozens of books written both specifically about budgies and including them across categories, age groups and concentration spans, ergo *Parakeets for Dummies* versus Jennifer Ackerman's *The Genius of Birds*. Children's books are particularly reflective of the times, ranging from the sweet innocence of *Parakeet Peter* (1954), the story of a little boy who receives a parakeet from his Aunt Susan and proceeds to teach it to talk, to the zombie bird, also called Peter, in the *Flight of the Pummeled Parakeet* (2015)—the sixth book in Sam Hay's Undead Pets series.

Their tweety image has been used to sell everything: whisky, waffle syrup, laundry detergent, home loans and even cigarettes. John Player's collectable cigarette trading cards included an aviary and cage bird series in 1933, with budgerigar varieties represented on four cards in the set of 50, while a magazine ad for Old Gold features one sitting on the fingers of a hand holding a lit cigarette.

Budgie owners also represented an incredibly lucrative market in the *Mad Men* era, with firms like American Bird Products offering everything from moulting food, blood tonic, bird wash, bird bitters, mating food and vitamins to boxes of fluff for nesting material. French's parakeet seed promised to keep birds lively and came with the added bonus of a 'pop-up vigor-building biscuit', toys, baths, and deluxe centrally heated cages with automated showers. The unfeathered kids weren't left out of this merchandising bonanza either, with such highly desirable items as a fourteen-piece parakeet circus kit that could only be obtained by sending in labels of three different parakeet food products—plus 50 cents.

But the *pièce de résistance* of marketing mojo reflected the explosion of products in the sanitary category. The tiny cotton-knit parakeet diaper or budgie nappy didn't exactly fly out the door when it was first offered for sale in the late 1950s. It took a couple of bird-loving naval officers from Virginia to really establish this market almost 40 years later by selling them as FeatherWear™ Flight-Suits™, complete with little disposable 'flight liners'. Their company Avian Fashions now dominates this niche, which allows birds to fly free 'without sacrificing the furniture'. The company, which employs at least fifteen people and has distributors in more than ten countries including Australia, has expanded its range to include leashes, hoodies, cable-knit jumpers, bandanas, hats and dress-up costumes (including pirate, naturally) in fifteen sizes, from 'Petite' for budgies to 'Colossal' for the largest macaws.

The budgie FlightSuit is not the most popular choice, but according to the folk at Avian Fashions this doesn't reflect actual bird numbers so much as *Melopsittacus*'s disdain for hotpants. 'The reason American budgie-sized (Petite) FlightSuits are not as popular is not necessarily because they are not popular pets, but because they are not readily handled,' a team member explains. 'If you try a budgie that is not tame, it takes a long time to tame it and subsequently condition it to wear a FlightSuit. The younger the bird when you start, the easier it is.'

Whether or not you think fashion is strictly for the birds, the budgies' lot has improved somewhat since the nineteenth century. Rather than giving their lives to adorn hats as they once did, they are now being repurposed post-mortem with a rather pointed message about the expendability of wildlife in the twenty-first century.

Emily Valentine is a highly awarded Sydney artist who has mastered the dead budgie as a medium since winning the World of WearableArt (WOW) design competition 'bizarre bra' category in 2002.

'I used to do fashion jewellery and I had been doing a lot using feathers, but the bra was my first piece of taxidermied wearable art,' Valentine recalls. The budgerigar brassiere came about as a result of a small domestic tragedy and became a means to celebrate Valentine's dead pets Rocky and Rolly.

'It was the classic story. I went away on holiday and left my birds with my friend and my friend forgot to bring them inside one after-noon, so they were both dead in the morning. When my friend rang up and told me the sorry story, I said, "Well, put them in the freezer." So, I had two dead budgies and I was aware of the competition so the two just came together like that. And it won.'

Since then Valentine has created many pieces of wearable art, as well as an extremely popular series of sculptures of dogs made from ethically sourced feathers of different birds. Whole birds some-times become available too. For example, one of her neighbour's budgies was killed in a close encounter with a cat, and Valentine deconstructed the bird into five brooches. Her *Sulphur Crested Frockatoo*, a cocktail dress and hat made from the heads and feathers of sulphur-crested cockatoos, won the major WOW Factor Award in 2014, against competitors from 40 countries.

Underlying the playful plumage is a broader wildlife message from Valentine, a committed member of WIRES (NSW Wildlife Information, Rescue and Education Service). 'In 2005 I curated an exhibition and I was travelling around Australia and I was just shocked at the roadkill, absolutely overcome by it. There is all this about the number of people that die on the road, but nothing about the other sentient beings. To me it sort of underscores how careless Australians are about roadkills.'

There is also growing awareness of the harsh reality of keeping birds in what often amounts to solitary confinement—denied the very birdiness of their nature. Even as far back as the 1930s there was recognition of the cost of too-close confinement of these creatures, with the RSCPA setting up 'flying schools' with large flights for birds that had never been allowed to stretch their wings.

In more recent times, the concept of animal dignity and sentience has been enshrined in laws in New Zealand, Europe, Canada and in the Australian Capital Territory. The sentience of living creatures is the nexus of the increasingly heated animal rights versus welfare debate, which is often brought into graphic focus by live export of larger animals like sheep and cattle. Now it is trickling down.

In Switzerland, for example, it is illegal to keep 'social species' including goldfish, guinea pigs and budgerigars in isolation. In Sweden, budgerigars must also live with at least one parent bird for the first six weeks after hatching and it is illegal to trim the wings of birds under one year old. In the Italian town of Reggio Emilia, the local authorities introduced a by-law requiring budgerigars to be kept in pairs and dictated that cages must be a minimum size of five times the bird's wingspan. These are the kinds of basic laws many would like to see introduced to cover small pets that have so far not been granted the attention given to larger animals.

Michelle McKee is one bird owner who hopes that in years to come, people will become sufficiently enlightened so that caged birds are no longer commodified and sold in pet shops. She and her husband, Harry, are better known to the budgie-loving community as the caretakers of a blue budgie called Cooper (of Cooper's Corner). San Diego-based Cooper is a star of social media with her own blog, Facebook profile and Instagram account, and she is gently trying to change perceptions as a little avian advocate.

'I think she is just so wonderful that I want to share her with the world. She has really changed our lives. When Harry got a job working outside the home we brought home Dewey, a little companion so she wouldn't be alone. Then it was like potato chips. You just can't have one,' says McKee.

'But we are against irresponsible breeding and as time has gone on, we don't think birds should be pets at all. We justify our having them only because we rescue. Our birds are either from the shelter or rehomed, and we do our best to give them a good life and spoil them. We have had eleven since we got Cooper. Three of our birds were found outdoors—one of them had been attacked by a cat.'

Because budgies are cheap to buy people often mistakenly believe they must be easy to keep. 'They don't realise how noisy, messy and expensive they can be. They need big cages, nutritious food (not just seed), lots of attention, toys, stimulation and companionship—especially if they are a single bird,' McKee explains.

'Birds you find in the big chain pet stores can be over-bred. Many will have health issues down the line—tumours or hormonal issues, for example. Females can lay themselves to death, become eggbound or have prolapses.' McKee says that over the years they have spent thousands of dollars at the vet. 'An avian vet can be hard to find and treatment can be very expensive. Not everyone can or wants to pay a vet bill in the hundreds, so they just let nature take its course.'

Cooper's message is adopt, don't shop. 'If you want a budgie check your local rescues and shelters. It would be lovely if a budgie wasn't even available to buy at a store. California has banned the sale of non-rescue dogs, cats and rabbits, but your little pocket pets—mice, rats, hamsters and little birds—they still just sell them like canned goods.'

That is some mountain of opinion to move. Over the 180 years since its first export, the budgerigar has several times run close to being the most popular pet in the world and even now retains its ranking as the most popular cage bird, and fourth-most popular household animal addition behind dogs, cats and fish.

But the halcyon days when people were content to listen to birds warbling on the wireless for entertainment are over. In Britain alone the budgie population plunged from 4 million in 1964 to 1.8 million by 1987. It is now estimated that only 1 per cent, or 270,000 British households, have 'indoor birds'. This had its inevitable effect on the birdy economy. The Scunthorpe Budgerigar Hotel, a novel bird-boarding business which over the years got lots of media attention, might not get off the ground today.

But as numbers have fallen in the traditional strongholds the budgie has attracted a new audience. Grant Findlay, administrator of the Budgerigar Society (UK) and a judge with the World Budgerigar Organisation, revealed new members, many of them aged under 30, flocking to the fancy in Bangladesh, Indonesia and Pakistan and across the Emirates. 'In the UK branch of the hobby we have just taken on over 40 members from Kuwait in the past few months. Worldwide it is growing, but in the traditional countries like the UK, Australia, USA, Germany and the like the hobby is at an even keel. We get as many new in each year as we lose.' But the internet is helping spread the word, Findlay notes. 'The hobby is very robust in Asia at the moment. They are like sponges out there and can't get enough information from us.'

In Japan, too, the budgie is enjoying a nostalgia-driven revival. Bird cafés, where customers can interact with budgies, other parrots and even owls are now a thing in Tokyo. One café even offers Sekisei Inko Ice (budgerigar-flavoured ice cream) which is basically vanilla ice cream with seeds and fruit. Budgie-themed merchandise is a growing category.

Clearly, we have not yet heard the end of the story of the little bird that grew!

Chapter 25

Wild thing

FROM CAPTURING THE silver-bell call of the critically endangered night parrot to stumbling across a rare woma python shining amber after just sloughing its skin, zoologist Mark Carter has experienced many wonders. Yet he regards 'budgie plagues' as the pinnacle.

'People talk about things that are quintessentially Australian or really big signifiers of the land and for me, those big budgie events are the peak of life in the outback. They always result from a boom season where there has been plenty of rain, plenty of resources, the right pattern of wet and dry to boost the numbers. So, when that happens it is almost like an expression of joy in the landscape.'

Even after being immersed in the glaring splendour of remote western and central Australia for more than a decade, the sight of a budgie murmuration never fails to thrill Carter, a bat and bird specialist who set up BirdLife Central Australia: 'The really, really big events are so kind of mercurial. They are very hard to pin down and predict.'

One of the most impressive budgie clouds Carter has ever seen was just outside Port Hedland, in the north of Western Australia.

'On the horizon I could see all this smoke,' he recalls. 'I thought it was a funny time of year for a bushfire. There had just been a cyclone and usually a fire can't get leverage in those conditions but as I got closer, I realised it was budgies. That's how many budgies there were—that they looked like the smoke from a giant wildfire.

'It is incredible how the energy from grass powers this big natural phenomenon that cascades through the food chain. It is a wonderful event to see unfold in front of you. The magnificence of nature is really in your face. No one can be left cold by it.'

Well, no one except some of the budgies themselves, who glisten like little green and yellow lollies to a range of predators including the grey falcon, an extremely rare and almost silver-coloured bird that numbers fewer than 1000 pairs in the wild. 'When you see those big plagues of budgies you start looking around for the grey falcon because they are pretty much budgie specialists who follow them around the outback,' Carter reveals.

 Power in numbers

Back in the late 1950s or early 1960s the budgerigars decided to use the powerlines east of Dalby (Darling Downs, Qld) to spend the long days. The wires were stretched with the weight and almost reached the ground. After the birds had moved on, the lines needed replacing as they had stretched so far. The flock filled the wires for several miles (as it was then) and the number of birds would have been incredible. .

—Terry Pacey, Birding-Aus forum, 1 May 2005

Meanwhile, in the Pilbara, the world's biggest predatory bat can be found hanging upside down in caves in the dark gorging itself on budgie sweetmeats just as it has since time immemorial. Carter vividly describes going into ghost bat roosts and finding piles, veritable middens, of budgie body parts. 'When those plagues are going on the bats are going out, grabbing the budgies off branches, bringing them back and just eating the liver and brain and pretty much throwing the rest of the budgie away. There were just these big piles of green and gold feathers.'

Huge swarms of budgerigars have also been recorded in more populous areas. Old timers still talk of the incredible migration of more than 1 million budgerigars that passed through the South Australian districts of Strathalbyn and Macclesfield in a single October day in 1963.

The late Peter Disher, birdwatcher and author from Barham on the New South Wales side of the Murray River, recalled the years when flocks of budgies darkened the sky from 'post office to butcher's shop' as they descended on the black box trees to the north of the town. They would swoop in to roost at such speeds that there were casualties, with Disher once collecting seven wings that had been sheared off as the flock came to land. The old trees where they once nested are now gone but for some roadside remnants. Geoff Leslie, a resident of Barham for twenty years, said it had become rare to see budgies in the area by the 1990s. 'In all my years I only saw them twice, the biggest flock consisting of thirteen birds.'

The world has little inkling about the ways of the wild budgie, which paradoxically is a victim of its own success in being able to thrive stupendously well in captivity. The European bird population by contrast is absurdly well studied. 'There are people there that can look at a seagull and tell you its precise age by the procession of moults and how dark a certain feather is on the wing,' Carter says.

'That sort of thing occurs in Australian birds too, but we haven't had anything like the number of people looking or the same tradition of gathering that data. And let's face it, the budgie is nowhere near as sexy as the other Australian parrots, like the rare ones that people go a bit gaga for.'

Nonetheless, to witness a budgie murmuration is on the tick list of every serious twitcher and visiting wildlife film crew, some of whom seem to think this event can be conjured to coincide with their arrival dates. 'It just doesn't work like that,' says Carter, who used to run bird tours. 'The few times people have pulled it off have been pure luck.'

When birds do mass at a waterhole, the experience is something like a green tornado—a deafening swoosh of upwards of 50,000 pairs of beating wings united by a single purpose. Witnesses are gobsmacked. 'You would have really seasoned birdwatchers who had been there, done that, travelled the world and seen 9000 species, but it melts people's hearts to see that event,' Carter says.

The wild creatures are a far cry from Bluey, the budgie of his youth spent in Scotland. Bluey almost certainly met an unhappy end after Carter's mother let him out of his cage for a fly and he went right out the window. 'It hung around our street in a pear tree for a bit, but we never got it back and I reckon it was eaten by something very quickly.' Carter recalled the lesson of his first bird when he saw a blue budgie, most likely a natural mutation, easily plucked from the mass by a goshawk.

Budgies of a different feather very seldom last long in the wild. 'You have to feel sorry for escaped budgies because they do try and make friends with the wild ones and the wild ones want nothing to do with them because they know they are a magnet for disaster,' he says.

'I am pretty convinced cage birds, when they get loose, don't really like life in the wild. It is scary and they spend most of their time hungry and they die quickly, so whenever people talk about it being cruel to keep them in a cage and want to let them go, I sort of point out that is a really bad idea.

'Wild birds have a really strong cultural life and social life which cage birds are never going to be able to replicate. I think they have been in captivity for so many generations now they are going to be chalk and cheese . . . They are just radically different.'

Warren Wilson is the only Australian recipient of the World Budgerigar Organisation's prestigious Gould Award. He's one of the few people who breeds all varieties of budgerigars, from the selectively bred show-bench birds, and the miniature or pet shop birds, through to their diminutive bush progenitors.

A former exhibition and conference convenor from Croydon in Sydney, Wilson counts no favourites among his flock, although he clearly has deep admiration for his 'bush budgies', whose lineages can be traced back to wild-caught birds from both New South Wales and South Australia. 'To my mind they are much smarter, although they have a very different nature, and good luck trying to get one to talk. The bush budgie does not want to talk to you. He has more important business on his mind.

'My bush budgerigar set-up is hardly touched. You don't have to worry about them because they look after themselves. They clean out their own nest boxes like you wouldn't believe—they strip everything out of them.

'It is incredible when you see the small holes they go into and raise so many chicks in there that sometimes they cannot get to the floor of the box and have to hang upside down from the nest hole to feed the babies. I made the mistake many years ago of thinking I would give them some nice big nest boxes—and they wouldn't have a bar of them.

'I have some zebra finches in the bush budgie aviary because in the wild that is who they bump into. They don't fight them. They don't chop their legs off like they do with canaries sometimes. No, there wouldn't be a one-legged zebra in the whole colony.'

Bush budgies may be small, but they are very feisty. 'If you get bitten by one you will know it,' Wilson insists. 'A bush budgie will inflict a bigger wound on you than a bigger bird. You have to understand where he is coming from and what he has to deal with in the wild. They get attacked by every goddamn thing in the bush.'

It's not just snakes slithering into the nest hole and eating the young that budgies must contend with. 'They need to worry about just about every other bird,' Wilson explains. 'They go into a little hole to breed and along comes a bigger parrot wanting the hole and you might think he'd win the argument, but it is not always the case with a bush budgie. He will put up such a fight that some of the other parrots won't take them on after the first 5 seconds.

'Make no mistake—they can be very, very vicious. If there are two blokes wanting the same hen there will be a stoush on you wouldn't believe. I've had to go in [the aviary] sometimes and separate them and literally take one away for fear one will kill the other. If two hens want the same nest box there will also be a real donnybrook.'

The picture of the scrappy little outback bar-room brawler is a far cry from the stagey, selectively bred exhibition bird or the playfully parlaying pet shop parrot. Wilson is one of an increasing number of fanciers who believe backyard geneticists are now pushing the species' envelope too far.

'Personally, I don't like the English budgerigar they are now promoting as being our top budgerigar. I think it is ugly and nothing like a normal budgerigar. Half of them don't fly and if they weren't locked up in an aviary would be dead in 5 seconds. It is a hopeless-looking thing and they shouldn't be promoting it to get any bigger,' Wilson says.

'If they grew that big naturally that is fine, but they don't. You have got to force them to grow that big and I don't think it is right. They end up with so much feather they can't fly. They get feather in the eye and can't fly because they can't see. Some of them won't even get off the floor of the cage. It is not what a bird is supposed to be like and that's not what the whole thing is supposed to be about.

'It would be alright if people stuck to the standard. The standard is what everybody has agreed the bird should look like. This going past the standard with ever bigger, fluffier birds is rubbish.'

More than 150 years of intensive select breeding, favouring the qualities that humans want to see rather than those that might serve the bird if they developed naturally, have had an inevitable effect on the creatures' physiology. They are shorter-lived, less fertile, more prone to disease, less adept at parenting and simply incapable of foraging for themselves.

'They used to be the hardiest birds of all time,' Wilson laments. 'They used to live ten, even fifteen years. Now if they reach six you are lucky. They are more susceptible to getting little things wrong with them. There must be 50 different medications for them now where they used to be five.'

The widening gap between the lithe, busy bushie and the couch-potato creature crafted through genetic selection has led some breeders to return to the original both for inspiration and assistance. Since capturing wild budgies is illegal in all states except South Australia, where they can be taken from the wild with a permit, this is often easier said than done.

Andrew McFarland, a commercial flower grower and champion budgie breeder, uses bush budgies as foster parents for his show birds' eggs and has sold them to others for this purpose. 'They are great with eggs and great feeders at that early stage. They work extremely hard because they are agile, small, robust and extremely vigorous to feed.

'Originally I got the green ones because I love the flock flying, the coordination they have together—it's like fish swimming in school. We have some nice big aviaries here that are 20 metres long. I love to go in there with 200 birds and see them fly past you in perfect unison, dipping and weaving.'

Almost as soon as McFarland established his colony of the wild types, he noticed the same mutations appearing that first excited early breeders in the 1860s and 1870s. He began reinventing the budgie anew and now counts golden face, cinnamon, opaline (a variety with fewer black markings on its head and shoulders), skyblues, goldenface mauves, violets and fallows (with red eyes, pink legs and feet and an orange beak) among his flock of bush budgies. These colours have been achieved without outcrossing from original green birds. 'These mutations occur naturally in the wild. It is just that they get banged off by hawks as soon as they appear in the flock,' the farmer explains.

The key difference between what McFarland is doing in his

breeding cabinets and what breeders of show varieties do is that he is not trying to selectively increase the birds' size or feather density. He is trying to strengthen the new mutations and hold the varieties and colours, but his birds remain feisty, busy little parrots true to their original phenotype.

Some people use bush budgies to 'liven up' the bigger birds in their aviaries with relative confidence they won't mate and spoil the show bloodlines. This is not to say a bush budgie cock won't try to mate with female budgezillas, but excess feather and the size differential make successful mating nigh on impossible.

Paul Hull has been fascinated to observe the contrast between bush budgies and his show birds. 'It's like comparing apples with oranges,' the Tamworth Budgerigar Society secretary says. 'The way they rear their chicks is totally different. In show birds you will have birds fledging a week apart whereas the natural wild birds would all fledge within a day, even though the eggs are laid two days apart. I guess it is a survival thing, so everyone is ready when it's time to move on. The bushies are more fertile, more virile and they don't miss. The first time we bred with them we got fifteen chicks out of three nests.

'It is clear we have destroyed some of the nature in the bush bird by making it an exhibition bird. Making it fat and lazy is probably a crude way to put it. An exhibition bird is basically built, the same way you breed a dog for different styles and characteristics. I guess you have to wonder if we have reached a line in the sand after over 100 years of selective breeding.'

The increasingly large, fluffy exhibition birds being promoted by some breeders in Europe particularly are, many believe, not so much budgies as cartoon versions of the real thing.

Photographer Steven Pearce is one of those lucky enough to have witnessed wild budgies at a seasonal zenith in 2012 and again in 2017. 'Nothing I have experienced with wildlife even comes close,' he says. 'This event is born of boom and bust environmental cycles, but I don't think even many Australians today would realise that it is a native wild bird that has been sent all over the world as a pet. If they do, they think it's just a little green parrot and we have many parrots.

'Times, I think, are changing. We have greater awareness about animals' intelligence and their ability to think, whereas before birds were just a thing that went chirp, chirp in the corner. You put a bird in a cage and clip its wings and it starts a trend, then another person picks it up and makes it bigger and so on over the course of a few generations until it gets to the point of "What the fuck have we done?" The point where you realise you have done a grotesque thing is always too late. That is human nature and maybe nature in general.

'You might say the natural selection process will have that same effect. For example, these murmurations grow up in this great boom period over several years and then, all of a sudden, one harsh year and they are all going to die.'

Perhaps the last word belongs to Warren Wilson, who was first introduced to the birds by his grandfather and even now, aged 82 himself, remains a tireless advocate. 'I think many people don't appreciate this is the home of the budgerigar,' he says. 'We call it Australia's international bird. It is on every continent except Antarctica and in most countries in the world. It is a phenomenal thing and people just do not realise how many there are and how important the budgerigar has become to the world pet industry.'

Acknowledgements

Without the generosity of friend and former colleague Robert Wainwright there would be no bumper book of budgies. Rob introduced us to publisher nonpareil Richard Walsh who, in turn, plucked a single green and gold feather of hope from the ashes of a story pitch made over coffee outside the old Castlemaine Fire Station late one afternoon in April 2018. This phoenix Richard nurtured towards full flight.

To tell the story of the budgerigar and its diaspora we were fortunate to enlist the aid of many experts and are particularly grateful to Walter Boles and Jacqueline Nguyen of the Australian Museum for their help on the paleontological period of the book.

The invaluable Australian brains trust of academics, vets and researchers also included Mark Carter, Rob Marshall, Peter Wilson, John de Nobile and Darryl Jones with Espen Odberg, Robert Dooling, Andrew Iwaniuk and Louis Lefebvre kindly assisting from abroad.

From the ranks of budgerigar breeders we thank Geoff Capes, Grant Findlay, Colin Flanagan, Henry George, Paul Hull, Andrew McFarland and Andy Pooley for their contributions.

Acknowledgements

We are most indebted to Alan Rowe, Vic Murray, Kelwyn Kakoschke and Warren Wilson who answered endless queries with their encyclopaedic knowledge of the bird business. Barney Enders, thanks for allowing use of your fabulous photograph of birds in the wild.

Birds we must credit for adding to the storyline include MacGyver, who was once famously lost in the wintery wilds of Winnipeg but rescued thanks to Shelley and Val Corvino and avian welfare advocate Melanie Shura.

A big spray of gold millet also goes to Cooper, a social media star with a message of ethical bird-keeping helpfully translated by her people Michelle and Harry McKee.

June Redmond, our secret first dear reader who encouraged us forward, would probably prefer champagne. Cheers to you June.

Author extraordinaire Sonya 'Ant' Hartnett provided insight into the publishing process and kindly put us in the way of some fabulous contacts.

Proving we really do have the best friends, another talented buddy Kylie Goldsack made posing for author headshots positively fun.

Allen & Unwin editorial director Rebecca Kaiser kept the flighty first-time authors from hyperventilating and made sure budgies were, at all times, in expert, professional hands. *Budgerigar* benefited hugely from the careful consideration of copy editor Emma Driver and the proof-reading prowess of Julia Cain.

Finally, the book was made to pop like the perkiest parrot on the planet with the cover design by Josh Durham with delightful, winsome little budgie motifs flying across the pages by Mika Tabata.

Our dog Smudge, it should be publicly noted, did absolutely nothing.

Bibliography

Books

Ackerman, J., *The Genius of Birds*, New York: Penguin Press, 2016.

Alderton, D., *Birds of the World*, London: Lorenz Books, 2018.

Arnold, K., Porter, R. & Wilkinson, L., *Animal Doctor: Birds and beasts in medical history*, London: Wellcome, 1994.

Ashley, M., *The Birdman's Wife*, Melbourne: Affirm Press, 2016.

Astley, H.D., *My Birds in Freedom and Captivity*, London: J.M. Dent & Co., 1900.

Baudin, N. (trans. C. Cornell), *The Journal of Post Captain Nicolas Baudin, Commander-in-Chief of the Corvettes Géographe and Naturaliste*, Adelaide: Libraries Board of South Australia, 1974.

Beeton, I.M., *Mrs Beeton's Book of Household Management*, Melbourne: Ward, Lock & Co., 1950.

Birkhead, T., *Bird Sense: What it's like to be a bird*, London: Bloomsbury Publishing, 2013.

—— *The Most Perfect Thing: Inside (and outside) a bird's egg*, London: Bloomsbury Publishing, 2017.

—— *The Wisdom of Birds: An illustrated history of ornithology*, London: Bloomsbury Publishing, 2008.

Biskind, P., *Star: The life and wild times of Warren Beatty*, London: Simon & Schuster, 2010.

Bodkin, F. & Robertson, L., *D'harawal: Dreaming stories*, Sussex Inlet, NSW: Envirobook, 2013.

Boehrer, B.T., *Parrot Culture: Our 2500-year-long fascination with the world's most talkative bird*, Philadelphia: University of Pennsylvania Press, 2010.

Bower, T., *Branson: Behind the mask*, London: Faber & Faber, 2001.

Boyde, M. (ed.), *Captured: The animal within culture*, London: Palgrave Macmillan, 2014.

Bransbury, J., *Where to Find Birds in Australia*, Adelaide: Waymark, 1998.

Branson, R., *Losing My Virginity*, London: Virgin Publishing, 1998.

Brendon, P., *Churchill's Bestiary: His life through animals*, London: Michael O'Mara Books, 2018.

Broinowski, G.J., *The Birds of Australia*, Melbourne: G.J. Broinowski, 1888.

Brown, A.G, *Ornithological Field Diaries*, Melbourne: Museums Victoria, 1957.

Brunner, B., *Birdmania*, Vancouver: Greystone Books, 2017.

Burton, R., *Bird Behaviour*, New York: Knopf, 1985.

Butler, A.G., *Foreign Birds for Cage and Aviary*, London: Feathered World, 1910.

Campbell, C., *Bonzo's War: Animals under fire 1939–1945*, London: Corsair, 2013.

Campbell, M., *Forget Not: The autobiography of Margaret, Duchess of Argyll*, London: W.H. Allen, 1975.

Carter, P., *Parrot*, London: Reaktion Books, 2016.

Cassell, J., *Cassell's Natural History: The feathered tribes*, London: John Cassell, 1854.

Cayley, N., *Australian Parrots: Their habits in the field and aviary*, Sydney: HarperCollins, 1938.

——*What Bird is That?* Sydney: Angus & Robertson, 1973.

Chambers, N. (ed.), *The Letters of Sir Joseph Banks: A selection, 1768–1820*, London: Imperial College Press, 2000.

Chhaya, M., *Dalai Lama: Man, monk, mystic*, Melbourne: Bolinda Publishing, 2008.

Chisholm, A.H., *Bird Wonders of Australia*, Sydney: Angus & Robertson, 1956.

—— *Mateship with Birds*, Melbourne: Scribe Publications, 2013.

Churchill, R.S. & Gilbert, M., *Winston S. Churchill*, vol. 3, Boston: Houghton Mifflin, 1966.

Clode, D., *Continent of Curiosities: A journey through Australian natural history*, Cambridge: Cambridge University Press, 2007.

Datta, A., *John Gould in Australia: Letters and drawings*, Melbourne: Melbourne University Press, 2002.

Delessert, E., *Souvenirs d'un Voyage à Sydney (Nouvelle-Hollande) fait pendant l'annee 1845*, Paris: Franck, 1847.

Doughty, R.W., *Feather Fashions and Bird Preservation: A study in nature protection*, Berkeley, CA: University of California Press, 1975.

Droscher, V.B., *They Love and Kill: Sex, sympathy and aggression in courtship and mating*, London: W.H. Allen, 1977.

Eastoe, J., *The Art of Taxidermy*, London: Pavilion, 2013.

Finn, F., *Ornithological and Other Oddities*, London: John Lane, 1907.

Flinders, M., *Terra Australis*, Melbourne: Text Publishing, 2012.

Forshaw, J.M., *Vanished and Vanishing Parrots: Profiling extinct and endangered species*, Melbourne: CSIRO Publishing, 2017.

Fowles, J., *Sydney in 1848*, Sydney: J. Fowles, 1878.

Fraser, I. & Gray, J., *Australian Bird Names: A complete guide*, Melbourne: CSIRO Publishing, 2013.

Frudd, P.G., *Funny Tales of Budgerigars Straight from the Author's Aviaries*, London: Read Books, 2011.

Gould, B., *Report on the Discovery, Recognition and Installation of the Fourteenth Dalai Lama*, New Delhi: Government of India Press, 1941.

Gould, J., *The Birds of Australia: In seven volumes*, London: John Gould, 1848.

—— *Handbook to the Birds of Australia*, London: John Gould, 1863.

——*The Mammals of Australia*, London: John Gould, 1863.

Greene, W.T., *Birds of the British Empire*, London: Imperial Press, 1898.

—— *Birds I Have Kept in Years Gone By*, London: L. Upcott Gill, 1885.

Greene, W.T., Dutton, F.G., Lydon A.F. & Fawcett, B., *Parrots in Captivity*, London: George Bell & Sons, 1883.

Griffiths, B., *Deep Time Dreaming*, Melbourne: Black Inc., 2018.

Hascall, V.C. & Balazs, E.A., *Karl Meyer 1899–1990: A biographical memoir*, Washington, DC: National Academy of Sciences, 2009.

Hodge, S., *Artists and their Pets*, Baltimore: Duopress, 2017.

Hoey, B., *Pets by Royal Appointment: The royal family and their animals*, London: Biteback Publishing, 2019.

Holt-White, R., *The Life and Letters of Gilbert White of Selbourne*, London: John Murray, 1901.

Howells, R., *Simply Churchill*, London: Robert Hale, 1965.

Hutton, D. & Connors, L., *A History of the Australian Environment Movement*, Melbourne: Cambridge University Press, 1999.

Isaacs, J., *Australian Dreaming: 40,000 years of Aboriginal history*, Sydney: Ure Smith, 1996.

Joseph, L. & Olsen, P., *Stray Feathers: Reflections on the structure, behaviour and evolution of birds*, Melbourne: CSIRO Publishing, 2011.

Jones, D., *The Birds at My Table*, Ithaca, NY: Comstock Publishing Associates, 2018.

Kass, D., *Educational Reform and Environmental Concern*, London: Routledge, 2019.

Kerr, J. (ed.), *Heritage: The national women's art book*, Sydney: Art & Australia, 1995.

Leach, J.A. & Lissenden, A., *An Australian Bird Book: A complete guide to the birds of Australia*, Melbourne: Whitcombe & Tombs, 1958.

Levi, P., *Edward Lear: A life*, London: Macmillan, 1995.

Low, R., *A Century of Parrots*, Mansfield, UK: Insignis Publications, 2006.

Low, T., *The New Nature: Winners and losers in wild Australia*, Melbourne: Penguin, 2002.

—— *Where Song Began: Australia's birds and how they changed the world*, Melbourne: Penguin, 2017.

Lubbock, J., *Animal Life and the World of Nature*, London: Hutchinson & Co., 1906.

Lydekker, R., *The New Natural History*, New York: Merrill & Baker, 1899.

Macinnis, P., *Curious Minds: The discoveries of Australian naturalists*, Canberra: National Library of Australia, 2012.

Mackay, C., *Extraordinary Popular Delusions and the Madness of Crowds*, New York: Farrar, Straus & Giroux, 1987.

Macpherson, E. ('Mrs Allan'), *My Experiences in Australia, Being Recollections of a Visit to the Australian Colonies in 1856/7*, London: J.F. Hope, 1860.

Marshall, R., *The Budgerigar*, Sydney: Rob Marshall, 2009.

Matthews, G.M., *The Birds of Australia*, London: Witherby, 1915.

Milgrom, M., *Still Life: Adventures in taxidermy*, New York: Mariner Books, 2011.

Mitchell, T., *Three Expeditions into the Interior of Eastern Australia*, London: T. and W. Boone, 1839.

Morris, E.E., *Austral English: A dictionary of Australian words, phrases and usages*, Cambridge: Cambridge University Press, 2011.

Nash, D., 'The smuggled budgie: Case study of an Australian loanblend', in R. Mailhammer (ed.), *Lexical and Structural Etymology: Beyond word histories*, Boston: Walter de Gruyter, 2013, pp. 293–311.

Neville, R., *A Rage for Curiosity: Visualising Australia 1788–1830*, Sydney: State Library of New South Wales Press, 1997.

Nightingale, F., *Notes on Nursing: What it is, and what it is not*, London: Harrison, 1859.

North, A.J., *Nests and Eggs of Birds Found Breeding in Australia and Tasmania*, Sydney: F.W. North, 1904.

North, M.R., *Everything About Parakeets (Budgies)*, New York: Hartz Mountain Products Corp., 1950.

Olsen, P., *Flocks of Colour*, Canberra: National Library of Australia, 2013.

Owen, R., *The Life of Richard Owen*, London: J. Murray, 1895.

Parker, K.L., *Australian Legendary Tales*, Melbourne: Melville, Mullen and Slade, 1896.

Parkinson, S., *A Journal of a Voyage to the South Seas in His Majesty's Ship the Endeavour*, London: Stanfield Parkinson, 1773.

Peck, M. & Samuel, D., *The Extraordinary Beauty of Birds*, Munich: Prestel, 2016.

Peck, R.M., *The Natural History of Edward Lear*, Boston: David R. Godine, 2016.

Pizzey, G. & Knight, F., *The Field Guide to the Birds of Australia*, Sydney: HarperCollins, 1980.

Quammen, D., *Spillover: Animal infection and the next human pandemic*, New York: W.W. Norton & Company, 2012.

Randolph, E., *The Basic Bird Book*, New York: Fawcett, 1989.

Rogers, C.H., *Encyclopedia of Cage and Aviary Birds*, London: Treasure Press, 1981.

Roth, W.E., *Ethnological Studies among the North-West-Central Queensland Aborigines*, Brisbane: Edmund Gregory, 1897.

Rowan, W., *The Riddle of Migration*, Baltimore: Williams & Wilkins, 1931.

Russ, K. (trans. M. Burgers), *The Budgerigar: Its natural history, breeding and management*, 7th ed., London: Cage Birds, 1928.

Russell, H.W.S, *Homing Budgerigars: Their care and management*, London: Cage Birds, 1953.

—— *Parrots and Parrot-like Birds*, Neptune City, NJ: T.F.H. Publications, 1969.

Russell, R., *The Business of Nature: John Gould and Australia*, Canberra: National Library of Australia, 2011.

Sands & Kenny, *Sands & Kenny's Commercial and General Sydney Directory for 1858–9*, Adelaide: Archival Digital Books Australasia, 2007 (1858).

Scoble, J., *The Complete Book of Budgerigars*, Sydney: Blandford Press, 1987.

Shaw, G., *The Naturalist's Miscellany: Or, coloured figures of natural objects; drawn and described immediately from nature*, London: Nodder & Co., 1789–1813.

Sinclair, I., *Hackney, That Rose-Red Empire: A confidential report*, London: Hamish Hamilton/Penguin, 2009.

Slater, P., Slater, P., Slater, R. & Elmer, S. *The Slater Field Guide to Australian Birds*, Sydney: New Holland Publishers, 2009.

Snow, J., *On the Mode of Communication of Cholera*, London: John Churchill, 1855.

Sparks, J. & Soper, T., *Parrots: A natural history*, Melbourne: Lothian, 1990.

Stap, D., *Birdsong: A natural history*, New York: Scribner, 2005.

Strahan, R., *Rare and Curious Specimens: An illustrated history of the Australian Museum, 1827–1979*, Sydney: Australian Museum, 1979.

Sturt, C., *Narrative of an Expedition into Central Australia*, London: T. & W. Boone, 1849.

Szasz, K., *Petishism: Pets and their People in the Western World*, New York: Holt, Rinehart and Winston, 1968.

Toft, C.A. & Wright, T.F., *Parrots of the Wild: A natural history of the world's most captivating birds*, Berkeley, CA: University of California Press, 2015.

Uglow, J., *Mr Lear: A life of art and nonsense*, New York: Farrar, Straus & Giroux, 2018.

Velten, H., *Beastly London: A history of animals in the city*, London: Reaktion Books, 2016.

Vitacco-Robles, G., *The Life, Times and Films of Marilyn Monroe*, vol. 2, Albany, GA: BearManor Media, 2014.

Watmough, W., *Colour Breeding in Budgerigars*, London: Cage Birds, 1950.

—— *The Cult of the Budgerigar*, London: Nimrod Book Services, 1936.

Watson, D., *A Single Tree: Voices from the bush*, Melbourne: Penguin, 2016.

Webb, C.S., *The Odyssey of an Animal Collector*, New York: Longmans, Green, 1954.

Westwood, B. & Moss, S., *Natural Histories: 25 extraordinary species that have changed our world*, London: John Murray, 2016.

Wheelwright, H.W., *Bush Wanderings of a Naturalist*, Melbourne: Oxford University Press, 1979.

White, B., *The Natural History and Antiquities of Selborne*, New York: G.P. Putnam, 1901.

Woolf, V., Nicolson, N. (ed.) & Banks, J.T. (ed.), *The Letters of Virginia Woolf*, London: Hogarth Press, 1980.

Journal articles, reports and theses

Arnold, K.E., Owens, I.P.F. & Marshall, N.J., 'Fluorescent signaling in parrots', *Science*, 2002, vol. 295, no. 5552, p. 92.

Ashley, M., 'Elizabeth Gould: A natural history', *The Lifted Brow*, 2015, no. 28, pp. 119–25.

—— 'Leaves of a diary: Searching for Elizabeth Gould in the archives of the Mitchell Library', *TEXT Journal*, 2013, vol. 17, no. 2.

Austad, S.N., 'Candidate bird species for use in aging research', *ILAR Journal*, 2011, vol. 52, no. 1, pp. 89–96.

Boles, W.E., 'A budgerigar *Melopsittacus undulatus* from the Pliocene of Riversleigh, north-western Queensland', *Emu*, 1998, vol. 98, no. 1, pp. 32–5.

Birkhead, T.R., Schulze-Hagen, K. & Palfner, G, 'The colour of birds: Hans Duncker, pioneer bird geneticist', *Journal für Ornithologie*, 2003, no. 144, pp. 253–70.

Brien, P. (trans. D. Reboussin), 'Jean-Marie Derscheid', *Biographie Nationale*, 1971, vol. 37, pp. 211–35.

Burnet, F.M., 'Enzootic psittacosis amongst wild Australian parrots', *Journal of Hygiene*, 1935, vol. 35, no. 3, pp. 412–20.

Chen, J., Zou, Y., Sun, Y.-H. & ten Cate, C., 'Problem-solving males become more attractive to female budgerigars', *Science*, 2019, vol. 363, no. 6423, pp. 166–7.

Chinner, D.W., 'Observations on the effect of increased rainfall on birdlife in central Australia', *South Australian Ornithologist*, 1977, no. 27, pp. 188–92.

Coleman, E. 'Memorable occurrences among budgerigars', *Victorian Naturalist*, 1947, vol. 64, pp. 97–101.

Comben, P., 'The emergence of Frederick Strange, naturalist', *Proceedings of the Royal Society of Queensland*, 2017, vol. 122, pp. 67–77.

Coote, A., 'Pray write me a list of species . . . that will pay me best', *History Australia*, 2014, vol. 11, no. 2, pp. 80–100.

—— 'Science, fashion, knowledge and imagination: Shopfront history in 19th-century Sydney', *Sydney Journal*, 2013, vol. 4, no. 1, pp. 1–18.

De Nobile, J., 'Working scientifically with budgerigars in the primary classroom', *Teaching Science*, 2013, vol. 59, no. 4, pp. 46–50.

Dent, M.L., 'Animal psychoacoustics', *Acoustics Today*, 2017, vol. 13, no. 3, pp. 19–26.

Franklin, A., 'Relating to birds in post-colonial Australia', *Kunapipi*, 2007, vol. 29, no. 2, pp. 102–25.

Gale, M.-A., 'Poor bugger whitefella got no Dreaming', PhD thesis, Adelaide: University of Adelaide, 2000.

Harrison, H. & Doyle, A., 'The electronic digital computer and Melopsittacus Undulatus', *INFORMS Journal on Applied Analytics*, 1984, vol. 14, no. 4, pp. 1–103.

Hasseltine, H.E., 'Some epidemiological aspects of psittacosis', *American Journal of Public Health and the Nation's Health*, 1932, vol. 22, no. 8, pp. 795–803.

Hennessy, P., 'Churchill and the premiership', *Transactions of the Royal Historical Society*, 2001, vol. 11, pp. 295–306.

Honigsbaum, M., 'In search of sick parrots: Karl Friedrich Meyer, disease detective', *The Lancet*, 2014, vol. 383, no. 9932, pp. 1880–1.

—— 'Tipping the balance: Karl Friedrich Meyer, latent infections, and the birth of modern ideas of disease ecology', *Journal of the History of Biology*, 2016, vol. 49, no. 2, pp. 261–309.

Iredale, T. & Whitley, G.P., 'John Roach and the budgerigar', *Australian Natural History*, 1962, vol. 14, no. 3, pp. 99–102.

—— 'John Roach, the budgerigar and the unfortunate officer', *Proceedings of the Royal Zoological Society of New South Wales*, 1970, vol. 89, pp. 36–9.

Kenneally, K.F., 'John Gould: Nature's illustrious illuminator', *Landscope*, 2004, vol. 20, no. 1, pp. 32–40.

Macintyre, J.J., *Papers of James J. Macintyre, 1834–1867*, National Library of Australia, MS 9176.

Magarey, A.T., 'Aborigines' water-quest in arid Australia', report no. 6, Australasian Association for the Advancement of Science, 1895, pp. 647–58.

Meggitt, M.J., 'Gadjari among the Walbiri Aborigines of Central Australia', *Oceania*, 1966, vol. 37, pp. 124–47.

Meyer, F.J., Bauer, P.C. & Costabel, U., 'Feather wreath lung: Chasing a dead bird', *European Respiratory Journal*, 1996, vol. 9, no. 6, pp. 1323–4.

Meyer, K.F., Eddie, B. & Stevens, I.M., 'Recent studies on psittacosis', *American Journal of Public Health*, 1935, vol. 25, no. 5, pp. 572–9.

Moravec, M., Striedter, G. & Burley, N., 'Assortative pairing based on contact call similarity in budgerigars', Melopsittacus undulatus, *Ethology*, 2006, vol. 112, pp. 1108–16.

Mugford, R.A. & M'Comisky, J.G., 'Some recent work on the psychotherapeutic value of caged birds with older people', in R.S. Anderson (ed.), *Pet Animals and Society*, London: Baillière Tindall, 1975, pp. 54–65.

Noonan, P., 'Sons of science: Remembering John Gould's martyred collectors', *Australasian Journal of Victorian Studies*, 2016, vol. 21, no. 1, pp. 28–42.

North, A.J., 'The destruction of native birds in New South Wales', *Records of the Australian Museum*, 1901, vol. 4, no. 1, pp. 17–21.

Nottebohn, F., 'The origins of vocal learning', *The American Naturalist*, 1972, vol. 106, no. 947, pp. 116–40.

Olsen, Penny, 'The flight of the budgerigar', *National Library Magazine*, 2011, vol. 3, no. 2, pp. 18–21.

Pranty, B., 'The budgerigar in Florida: Rise and fall of an exotic psittacid', *North American Birds*, 2001, vol. 55, no. 4, pp. 389–97.

Robb, S.S., Boyd, N. & Pristash, C.L., 'A wine bottle, plant, and puppy. Catalysts for social behavior', *Journal of Gerontological Nursing*, 1980, vol. 6, no. 12, pp. 721–8.

Schiffner, I., Perez, T. & Srinivasan, M.V., 'Strategies for pre-emptive mid-air collision avoidance in budgerigars', *PLOS ONE*, 2016, vol. 11, no. 9.

Senate Standing Committee on Rural & Regional Affairs & Transport, *Australia's Biosecurity and Quarantine Arrangements*, Canberra: Department of the Senate, 2012.

Simberloff, D. & Gibbons, L., 'Now you see them, now you don't! Population crashes of established and introduced species', *Biological Invasions*, 2004, vol. 6, no. 2, pp. 161–72.

Stamps, J., Clark, A., Arrowood, P. & Kus, B., 'Parent–offspring conflict in budgerigars', *Behaviour*, 1985, vol. 94, no. 1–2, pp. 1–40.

Stapylton, G.W.C., 'Extracts from his journal', in M.H. Douglas & L. O'Brien (eds), *The Natural History of Western Victoria: Proceedings of the symposium*, Australian Institute of Agricultural Science, 1974, pp. 85–116.

Sutton, D., 'Maurice Baring: Sophistication and sentiment', *Apollo*, 1982, vol. 115, no. 243, pp. 394–400.

Thomson, D.F., 'The Bindibu expedition: Exploration among the desert Aborigines of Western Australia', *The Geographic Journal*, 1962, vol. 128, no. 3, pp. 262–78.

Ung, D., Amy, M. & Leboucher, G., 'Heaven it's my wife! Male canaries conceal extra-pair courtships but increase aggressions when their mate watches', *PLOS ONE*, 2011, vol. 6, no. 8.

Wenner, A.S. & Hirth, D.H., 'Status of the feral budgerigar in Florida', *Journal of Field Ornithology*, 1984, vol. 55, no. 2, pp. 214–19.

Whitehead, P.J.P., 'Zoological specimens from Captain Cook's voyages', *Journal of the Society for the Bibliography of Natural History*, 1969, vol. 5, no. 3, pp. 161–201.

Newspapers

Australia: *Adelaide Advertiser*, (Adelaide) *Chronicle*, *Adelaide Evening Journal*, *Adelaide Express and Telegraph*, *Adelaide Mail*, (Adelaide) *News*, *Adelaide Observer*, (Adelaide) *Register*, *Armidale Express and New England Advertiser*, *Australasian Chronicle*, *Bell's Life in Sydney and Sporting Reviewer*, *Braidwood Dispatch and Mining Journal*, *Brisbane Telegraph*, (Broken Hill) *Barrier Miner*, *Burnie Advocate*, *Cairns Post*, *Canberra Times*, *Corowa Free Press*, (Gawler) *Bunyip*, *Illustrated Sydney News and New South Wales Agriculturalist and Grazier*, *Maitland Daily Mercury*, (Melbourne) *Age*, (Melbourne) *Herald*, (Melbourne) *Weekly Times*, *Newcastle Sun*, *Nowra Leader*,

(Perth) *Sunday Times*, (Perth) *West Australian*, (Rockhampton) *Evening News*, *South Australian Chronicle and Weekly Mail*, *South Australian Register*, *South Australian Weekly Chronicle*, *Sydney Gazette*, *Sydney Morning Herald*, (Sydney) *Australian*, (Sydney) *Daily Telegraph*, *Sydney Directory*, (Sydney) *Sun*, (Sydney) *Truth*.

UK: *Belfast News*, *Belfast News-Letter*, *Belfast Telegraph*, *Birmingham Daily Gazette*, *Birmingham Daily Post*, (Birmingham) *Evening Despatch*, *Central Somerset Gazette*, *Cheshire Observer*, *Coventry Evening Telegraph*, (Dublin) *Irish Times*, *Dundee Evening Telegraph*, *Greenock Advertiser*, *Guardian*, *Northern Daily Mail*, *Hull Daily Mail*, *Liverpool Daily Post*, *Liverpool Echo*, (London) *Daily Mail*, (London) *Daily Mirror*, *London Daily News*, *London Evening Standard*, (London) *Financial Times*, (London) *Leader*, (London) *Observer*, (London) *People*, *Manchester Guardian*, *Scotsman*, *Sheffield Weekly Telegraph*, *Sunderland Daily Echo and Shipping Gazette*, *Sussex Agricultural Express*, *Times*, *Warwick & Warwickshire Advertiser and Leamington Gazette*, *Western Morning News*, *Yorkshire Evening Post*, *Yorkshire Post and Leeds Intelligencer*.

North America: (Bergen County, New Jersey) *Record*, *Boston Globe*, *Calgary Herald*, *Los Angeles Times*, (New Jersey) *Record*, *New York Times*, *Orange County Register*, *Ottawa Journal*, *Pittsburgh Sun-Telegraph*, *Rapid City Journal*, *San Francisco Examiner*, *San Jose News*, (Saskatoon) *StarPhoenix*, *Tampa Tribune*, (Toronto) *Globe and Mail*, *Tucson Daily Citizen*, (Victoria, BC) *Daily Colonist*, *Wall Street Journal*, *Washington Post*, *Winnipeg Free Press*, *Winnipeg Tribune*.

Magazines, journals

Australia: *Amytornis: Western Australian Journal of Ornithology*, *Australian Birdlife*, *Australian Bird Watcher*, *Australian Field*

Ornithology, Australian Geographic, Australian Museum Magazine, Australian Natural History, Australian Veterinary Journal, Australian Walkabout, Australian Wildlife, Australian Women's Weekly, Australian Zoologist, Bell's Life in Sydney and Sporting Reviewer, Bulletin, Emu: Austral Ornithology, Journal of the Historical Society of South Australia, Memoirs of the National Museum of Victoria, Northern Territory Naturalist, Oceania, Postcolonial Studies, Proceedings of the Linnean Society of NSW, Proceedings of the Royal Zoological Society of NSW, Records of the South Australian Museum, South Australian Naturalist, South Australian Ornithologist, Smith's Weekly, Tasmanian Naturalist, Victorian Naturalist, Wild Life, Wingspan.

UK and Europe: *Avicultural Magazine, Bird Notes, Bird World, British Association for the Advancement of Science Reports, British Birds, British Medical Journal, British Ornithologists' Club Bulletin, Cage Bird and Bird World, Country Life, Field, Hamlyn's Menagerie Magazine, Hardwicke's Science Gossip, Ibis, Interior Design, Knight's Penny Magazine, Lancet, L'Oiseau, Naturalist's Miscellany, Nature, New Scientist, Oriental Bird Club Bulletin, Ornis, Proceedings of the Zoological Society of London, PsittaScene, Sport and Country.*

North America: *Bird Lore, Journal of Avian Medicine and Surgery, Maclean's, New Yorker, Smithsonian, Wilson Bulletin.*

Websites

Angel Scrolls Forum: http://theangelscrolls.freeforums.net
Ausbird—The Australian Birding Directory: www.ausbird.com
Australian Bird Study Association: www.absa.asn.au
Australian Budgerigar Society: www.absbudgieclub.org.au
Australian Dictionary of Biography: http://adb.anu.edu.au

Australian Museum: https://australianmuseum.net.au

Australian National Budgerigar Council: www.anbc.iinet.net.au

Biodiversity Heritage Library: www.biodiversitylibrary.org

Bird Health with Dr Rob Marshall: www.birdhealth.com.au

Birding-Aus, mailing list archive: http://bioacoustics.cse.unsw.edu.au/
archives/html/birding-aus/

Birdlife Australia, 'Budgerigar': http://birdlife.org.au/bird-profile/
budgerigar

Birds Online: Everything About Budgies: www.birds-online.de/
allgemein/geschlecht_en.htm

Birds Queensland: The Queensland Ornithological Society:
www.birdsqueensland.org.au

Birds SA: https://birdssa.asn.au

Budgerigar Council of Tasmania: www.bctas.info

Budgerigar Council of Victoria: www.bcv.asn.au

Budgerigar 'Rare' and Specialist Exhibitors of Australasia:
www.brasea.com

Budgerigar Society, 'Society history': www.budgerigarsociety.com/
society-history/

Budgerigar Society of NSW: www.budgerigar.com.au

Budgerigar Society of South Australia: www.bssainc.org.au

Budgerigarworld.com: www.budgerigarworld.com

Budgies are Awesome: http://budgiesareawesome.blogspot.com

Currumbin Valley Veterinary Services: www.currumbinvetservices.com.au

Encyclopedia of Australian Science: www.eoas.info

Forbes: www.forbes.com

Japingka Aboriginal Art: https://japingkaaboriginalart.com

Library of Congress: https://loc.gov

National Audubon Society: www.audubon.org

Budgerigar

National Library of Australia Magazine: www.nla.gov.au/unbound

Omlet, 'The world's most popular pet bird': www.omlet.co.uk/guide/ budgie_guide/introduction_to_budgies/most_popular_bird/

Oxford Word of the Month: http://slll.cass.anu.edu.au/centres/andc/ oxford-word-month

Pet Directory: www.petdirectory.com.au

Pet Histories: The AHRC Pets and Family Life Project Blog: https://pethistories.wordpress.com/about

Pets and Their Authors: http://petsandauthors.blogspot.com.au

Presidential Pet Museum: www.presidentialpetmuseum.com

South Queensland Budgerigar Breeders Association: www.sqbba.com

Talk Budgies: www.talkbudgies.com

Wikipedia: https://en.wikipedia.org

Wikivisually: https://wikivisually.com